Lecture Notes in Mathematics

Edited by A. Dold, Heidelberg and B. Eckmann, Zürich

361

John W. Schutz
Monash University, Clayton, Victoria/Australia

Foundations of Special Relativity: Kinematic Axioms for Minkowski Space-Time

Springer-Verlag
Berlin · Heidelberg · New York 1973

AMS Subject Classifications (1970): Primary: 70 A 05, 83 A 05, 83 F 05
Secondary: 50-00, 50 A 05, 50 A 10, 50 C 05, 50 D 20, 53 C 70

ISBN 3-540-06591-1 Springer-Verlag Berlin · Heidelberg · New York
ISBN 0-387-06591-1 Springer-Verlag New York · Heidelberg · Berlin

Offsetdruck: Julius Beltz, Hemsbach/Bergstr.

To Amina

PREFACE

The aim of this monograph is to give an axiomatic development of Einstein's theory of special relativity from axioms which describe intuitive concepts concerning the kinematic behaviour of inertial particles and light signals.

I am grateful to Professor G. Szekeres and Dr. E.D. Fackerell for their encouragement and constructive suggestions during the preparation of this monograph.

John W. Schutz
Monash University

TABLE OF CONTENTS

CHAPTER 1. INTRODUCTION 1

CHAPTER 2. KINEMATIC AXIOMS FOR MINKOWSKI SPACE-TIME 7

2.1 Primitive Notions 7
2.2 Existence of Signal Functions 8
2.3 The Temporal Order Relation 9
2.4 The Triangle Inequality 12
2.5 Signal Functions are Order-Preserving 13
2.6 The Coincidence Relation. Events 14
2.7 Optical Lines 17
2.8 Axiom of the Intermediate Particle 24
2.9 The Isotropy of SPRAYs 25
2.10 The Axiom of Dimension 33
2.11 The Axiom of Incidence 35
2.12 The Axiom of Connectedness 36
2.13 Compactness of Bounded sub-SPRAYs 38

CHAPTER 3. CONDITIONALLY COMPLETE PARTICLES 42

3.1 Conditional Completion of a Particle 42
3.2 Properties of Extended Signal Relations and Functions 44
3.3 Generalised Triangle Inequalities 47
3.4 Particles Do Not Have First or Last Instants 48
3.5 Events at Which Distinct Particles Coincide 50
3.6 Generalised Temporal Order. Relations on the Set of Events. Observers. 52
3.7 Each Particle is Dense in Itself 57

CHAPTER 4. IMPLICATIONS OF COLLINEARITY 59

4.1 Collinearity. The Two Sides of an Event 59
4.2 The Intermediate Instant Theorem 62
4.3 Modified Signal Functions and Modified Record Functions 66
4.4 Betweenness Relation for n Particles 69

CHAPTER 5. COLLINEAR SUB-SPRAYS AFTER COINCIDENCE 71

5.1 Collinearity of the Limit Particle 72
5.2 The Set of Intermediate Particles 77
5.3 Mid-Way and Reflected Particles 84
5.4 All Instants are Ordinary Instants 95
5.5 Properties of Collinear Sub-SPRAYs After Coincidence 100

CHAPTER 6. COLLINEAR PARTICLES 103

6.1 Basic Theorems 103
6.2 The Crossing Theorem 119
6.3 Collinearity of Three Particles. Properties of Collinear sub-SPRAYs. 123
6.4 Properties of Collinear Sets of Particles 132

CHAPTER 7. THEORY OF PARALLELS 147

7.1 Divergent and Convergent Parallels 148
7.2 The Parallel Relations are Equivalence Relations 164
7.3 Coordinates on a Collinear Set 172
7.4 Isomorphisms of a Collinear Set of Particles 191
7.5 Linearity of Modified Signal Functions 210

CHAPTER 8. ONE-DIMENSIONAL KINEMATICS 233

8.1 Rapidity is a Natural Measure for Speed 233
8.2 Congruence of a Collinear Set of Particles 239
8.3 Partitioning a Collinear Set of Particles into Synchronous Equivalence Classes 243
8.4 Coordinate Frames in a Collinear Set. 246

CHAPTER 9. THREE-DIMENSIONAL KINEMATICS 250

9.1 Each 3-SPRAY is a 3-Dimensional Hyperbolic Space 251
9.2 Transformations of Homogeneous Coordinates in Three-Dimensional Hyperbolic Space 256
9.3 Space-Time Coordinates Within the Light Cone 263
9.4 Properties of Position Space 271
9.5 Existence of Coordinate Frames 278
9.6 Homogeneous Transformations of Space-Time Coordinates 286
9.7 Minkowski Space-Time 290

CHAPTER 10. CONCLUDING REMARKS 300

APPENDIX 1. CHARACTERISATION OF THE ELEMENTARY SPACES 302

APPENDIX 2. HOMOGENEOUS COORDINATES IN HYPERBOLIC AND EUCLIDEAN SPACES 309

BIBLIOGRAPHY 312

CHAPTER 0

SUMMARY

Minkowski space-time is developed in terms of undefined elements called "particles" and a single undefined relation, the "signal relation". Particles correspond physically to "inertial particles" and the signal relation corresponds to "light signals". The undefined basis is similar to that of Walker [1948].

Altogether there are eleven axioms. The first five are similar in content to those of Walker [1948]. Of the remaining axioms, four concern sets of particles which coincide at any one given event and which are called SPRAYs . We postulate:

(i) between any two distinct particles of a SPRAY, there is a particle which is distinct from both,

(ii) each SPRAY is isotropic,

(iii) there is a SPRAY which has a maximal symmetric sub-SPRAY of four distinct particles, and

(iv) each bounded infinite sub-SPRAY is compact.

The essential content of the remaining two axioms is that: space-time can be "connected" by particles; and that, given any two distinct particles which coincide at some event, there is a third distinct particle which forms the third side of a "triangle".

The ensuing discussion falls naturally into two parts; the

development of rectilinear kinematics, which is in many ways similar to the geometry of coplanar subsets in absolute geometry; and the extension to three-dimensional kinematics which is established by first showing that each SPRAY is a three-dimensional hyperbolic space and then extending a correspondence between homogeneous coordinates and space-time coordinates. These ideas will now be described in more detail.

Collinear sub-SPRAYs are shown to exist and their properties are discussed. Then the existence of maximal collinear sets of particles is demonstrated and it is found that they have many properties which are analogous to properties of coplanar subsets in the theory of absolute geometry. The concept of parallelism is applied to particles and, as in absolute geometry, we are faced with the possibilities of there being none, one, or two distinct particles which are parallel to a given particle through a given event. It is shown that there are two types of parallels, which may or may not be distinct, and that both types of parallels lead to equivalence relations of parallelism. The set of all events in a maximal collinear set can then be "coordinatised" with respect to any equivalence class of parallels. Both relations of parallelism turn out to be invariant with respect to reflection mappings. By composing several reflection mappings it is possible to generate "pseudo-rotations", space translations and time translations. It transpires that the uniqueness of parallelism is a theorem, which is a marked contrast with the theory of absolute geometry! This remarkable finding implies that each particle moves with uniform velocity.

It is shown that each SPRAY is a three-dimensional hyperbolic space, with particles corresponding to "points" and with relative velocity as a metric function. Homogeneous coordinates in three-dimensional hyperbolic space correspond to space-time coordinates "within a light cone". The extension of this correspondence to all events gives rise to the concept of a coordinate frame. Associated with each coordinate frame is a position-space which is shown to be a three-dimensional euclidean space. Transformations of homogeneous coordinate systems then correspond to "homogeneous Lorentz transformations" from which the "inhomogeneous Lorentz transformations" are derived.

§0.1. GLOSSARY OF DEFINITIONS AND NOTATION

This listing contains definitions and symbols in their order of appearance within the text. Numbers on the left refer to the number of the section in which the definition appears. The symbol □ indicates the end of a proof.

Section	Definition and Notation
2.1	particles $\underset{\sim}{Q}, \underset{\sim}{R}, \underset{\sim}{S}, \underset{\sim}{T}, \underset{\sim}{U}, \underset{\sim}{V}, \underset{\sim}{W}$
2.1	set of particles $\mathcal{P}$
2.1	instants $Q_a, R_1, S_x, \cdots$ (see also §3.1)
2.1	set of instants $\mathcal{I}$
2.1	signal relation σ (see also §3.2, §3.6)
2.2	signal function (from $\underset{\sim}{Q}$ to $\underset{\sim}{R}$) f_{RQ} (see also §3.2, §3.6)
2.3	record function (of $\underset{\sim}{R}$ relative to $\underset{\sim}{Q}$) $f_{QR} \circ f_{RQ}$
2.3	distinct instants $\neq$
2.3	temporal (order) relation(s) $<, >, \leqslant, \geqslant$ (see also §3.2, §3.6)
2.3	before-after $<, >, \leqslant, \geqslant$
2.4	direct signal
2.4	indirect signal
2.6	coincidence relation $\simeq$
2.6	event [ ]
2.6	set of events $\mathcal{E}$
2.7	optical line, in optical line | , , > (see also §3.2, §3.6)

Section	Definition and Notation
2.7	exterior to
2.8	permanently coincident particles $\simeq$
2.8	distinct particles $\neq$
2.9	between
2.9	betweenness relation for particles $<\ ,\ ,\ >$
	(also §3.2, §4.4)
2.9	SPRAY *SPR*[ ]
2.9	spray *spr*[ ]
2.9	isotropy mapping θ,ϕ,ψ
2.10	symmetric sub-SPRAY
2.12	connected (set of instants)
2.13	bounded sub-SPRAY
2.13	cluster particle
3.1	first (instant)
3.1	last (instant)
3.1	cut
3.1	gap
3.1	conditionally complete particle $\underset{\sim}{\overline{Q}},\cdots$
3.1	ordinary instant $Q_x, R_1, \cdots$
3.1	ideal instant $\overline{Q}_x, \overline{R}_1, \cdots$
3.1	instant $Q_c, \overline{R}_1, \cdots.$ (see also §2.1)
3.1	set of conditionally complete particles $\overline{\mathcal{P}}$
3.1	set of instants $\overline{\mathcal{I}}$ (see also §2.1)
3.2	extended signal relation (see also §2.1, §3.6)

Section	Definition and Notation
3.2	extended temporal order relation (see also §2.3, §3.6)
3.2	extended signal function (see also §2.2, §3.6)
3.2	extended coincidence relation
3.2	ideal event
3.2	extended relation of "in optical line" (see also §2.7, §3.6)
3.2	extended betweenness relation
3.6	generalised temporal order relation (see also §2.3, §3.2)
3.6	generalised signal relation (see also §2.1, §3.1)
3.6	generalised signal functions (see also §2.2, §3.2)
3.6	generalised relation of "in optical line" (see also §2.7, §3.2)
3.6	observer $\hat{\underset{\sim}{Q}}, \hat{\underset{\sim}{R}}, \hat{\underset{\sim}{S}}, \cdots\cdots$
3.7	dense
4.1	collinear particles $[\underset{\sim}{Q}, \underset{\sim}{R}, \underset{\sim}{S}, \cdots]$ (see also §4.4)
4.1	side (left, right) (see also §4.2, §6.2)
4.2	side (left, right) (see also §4.1, §6.2)
4.2	right optical line, left optical line
4.2	to the right of, to the left of
4.3	modified record function $\left\{ f_{QR} \circ f_{RQ} \right\}^*$
4.3	modified signal functions f^+, f^-

Section	Definition and Notation
4.4	betweenness relation after (before) an event < , , > *after* [ ]
4.4	collinear after (before) an event [, ,] *after* [ ] (see also §4.1)
5.1	limit particle
5.3	mid-way between (see also §7.1)
5.3	reflection (see also §7.1)
5.3	reflected observer $\hat{\underset{\sim}{Q}}_T$ (see also §7.1)
5.5	collinear sub-SPRAY *CSP*[ ]
5.5	collinear sub-spray *csp*[ ]
6.2	side (left, right) (see also §4.1, §4.2)
6.2	cross (verb)
6.4	collinear set (of particles) *COL*[ ]
6.4	collinear set (of events) *col*[ ]
7.1	parallel (divergent or convergent)
7.1	diverge from
7.1	converge to
7.1	reflection (see also §5.3)
7.1	reflected particle (see also §5.3)
7.1	reflected event
7.1	mid-way parallel
7.3	dyadic numbers, parallels, instants
7.3	indexed class of parallels
7.3	time scale (divergent, convergent)

Section	Definition and Notation
7.4	pseudo-rotation
7.4	spacelike translation (see also §7.5)
7.4	time translation (see also §7.5)
7.5	natural time scale (see also §7.3)
7.5	space displacement mapping (see also §7.4)
7.5	time displacement mapping (see also §7.4)
7.5	time reversed
8.1	constant of the motion
8.1	rapidity (directed, relative)
8.2	congruent particles
8.2	distance (directed)
8.3	synchronous particles
8.4	position-time coordinates
8.4	coordinate frame (in *coll*)
8.4	origin in position-time
8.4	origin in position
8.4	velocity
9.1	3-SPRAY *3SP*[ ]
9.2	origin of homogeneous coordinate system
9.2	homogeneous coordinates (see also §A.2)
9.3	space-time coordinates
9.3	origin in space-time
9.3	time coordinate
9.3	space coordinates

Section	Definition and Notation
9.3	origin in space
9.3	coordinate time identification mapping
9.3	within the light cone
9.3	light cone (upper, lower)
9.3	vertex
9.3	position space
9.4	parallel position spaces
9.5	coordinate frame
9.5	time coordinate transformation
9.6	homogeneous Lorentz transformation
A.1	topology
A.1	open sets
A.1	points
A.1	closed set
A.1	neighbourhood
A.1	closure
A.1	connected
A.1	open cover
A.1	subcover
A.1	compact space
A.1	locally compact space
A.1	metric space
A.1	distance
A.1	diameter
A.1	bounded

Section	Definition and Notation
A.1	curve
A.1	length of curve
A.1	arcwise connected
A.1	intrinsic metric
A.1	motion
A.1	doubly transitive
A.1	isotropic space
A.2	point
A.2	projective coordinates
A.2	special projective coordinates
A.2	change of basis

CHAPTER 1

INTRODUCTION

Following Einstein's formulation of the theory of special relativity (Einstein [1905]), several axiomatic systems have been proposed for Minkowski space-time. There are several reasons for developing a physical theory along axiomatic lines. One reason, which is not always made explicit, is a desire that special relativity may be better understood and more widely accepted. Another reason is that, if an axiomatic system is successful in clarifying concepts and exhibiting a small number of intuitively based assumptions, it is conceivable that some modification to one or more of the axioms might lead to an alternative theory of physical interest. Such was the case with the axiom system which Euclid [~ 300 B.C.] proposed for elementary geometry: subsequently Bolyai [1832] and Lobachevsky [1829] altered the Euclidean axiom of parallelism and discovered hyperbolic geometry! Similarly, we might expect that modification of an axiom system for Minkowski space-time could lead to a previously unknown, and possibly non-Riemannian, space-time of physical interest; a possibility which the reader is encouraged to keep in mind! (An example of a non-Minkowskian, but Riemannian, space-time which satisfies appropriately modified versions of the axioms given here, is the de-Sitter universe. The principal modification made is to Axiom I (§2.2)).

Prior to the present treatment, several axiomatic systems have already been formulated. Some authors assume the concept of a coordinate frame; in particular, Bunge [1967] has axiomatised the conventional approach due to Einstein [1905], while Suppes [1954, 1959] and Noll [1964] have based their systems on the assumption of the invariance of the quadratic form

$$\Delta x_1^2 + \Delta x_2^2 + \Delta x_3^2 - c^2 \Delta t^2$$

with respect to transformations between coordinate frames. Zeeman [1964] has shown that the inhomogeneous Lorentz group is the largest group of automorphisms of Minkowski space-time. Robb [1936] formulated an axiomatic system in terms of a single relation (before-after) between the undefined elements which he called "events", however Robb's aim of mathematical simplicity is achieved by selecting his axioms on mathematical rather than physical grounds. A system of axioms has been proposed by Walker [1948, 1959], who suggested foundations for relativistic cosmology in terms of an undefined basis which involved the concepts of "particles", "light signals" and "temporal order". Walker's axiom system was not developed sufficiently to describe Minkowski space-time and was, in fact, restricted to sets of relatively stationary particles, but it succeeded in clarifying many kinematic concepts, especially those of "particle", light signal", "optical line" and "collinearity". The undefined basis of Szekeres [1968] bears some resemblance to that of Walker, although Szekeres regards both particles and light signals as objects whereas Walker regards particles as objects and light signals as particular instances of a binary "signal relation". Of these three axiomatic systems, only that of Szekeres [1968] succeeds in describing Minkowski space-time in terms of assumptions related to what one might describe as either "kinematic experience" or physical intuition.

Our intention is to describe Minkowski space-time in terms of undefined elements called "particles", a single undefined relation called the "signal relation" and eleven axioms which are intended to be in accordance with the reader's physical intuition. In a subjective sense, a particle corresponds to a freely moving observer who is capable of distinguishing between "local" events; the concept of a particle can therefore be regarded as a more basic concept than that of a "coordinate frame" (which distinguishes between different events by assigning different sets of coordinates to them): the undefined signal relation corresponds physically to "light signals". This undefined basis is similar to that of Szekeres [1968] in a physical sense and to that of Walker [1948, 1959] in a formal sense, although whereas Walker [1948, 1959] used *two* undefined relations, the "signal relation" and the "temporal order relation", the present treatment has only *one* undefined relation, the "signal relation". The notion of time ordering is implicit in the concept of the signal relation and so temporal order can be defined in terms of the signal relation. Apart from this change and certain other modifications which result in a weaker set of assumptions, the first five axioms of the present system, together with their elementary consequences, bear a strong resemblance to the excellent analysis of the concepts of "particles", "light signals" and "collinearity" given by Walker [1948]. The subsequent six axioms are essentially different from those of Walker [1948, 1959] and are believed to be original in their application to special relativity. Four of these axioms refer to sets of particles which represent "velocity space"; they resemble axioms which have been used in the study of metric geometry by authors such as Busemann [1955].

Before stating the axioms it may be as well to point out that there are several assumptions which we do not make. In particular, we do not assume the concept of a coordinate frame, we do not assume that the set of instants of each particle can be ordered by the real numbers, nor do we assume that particles and light signals move with constant speed. In the present axiomatic system, these propositions turn out to be theorems.

Three properties of Minkowski space-time are of central importance to the subsequent development. One-dimensional kinematics is in many ways analogous to plane absolute geometry, for it transpires that the concept of parallelism can be applied to particles and, furthermore, the corresponding question of uniqueness of parallelism is closely related to the uniform motion of particles. Both Robb [1936] and Szekeres [1968] observed that uniform motion implies uniqueness of parallelism but, in the present treatment, we are able to *prove* the uniqueness of parallelism and then to show that this implies the uniform motion of (freely-moving) particles, so that we need not assume Newton's first law of motion explicitly. The second important property is that, in contrast to the euclidean velocity space of Newtonian kinematics, the velocity space associated with Minkowski space-time is hyperbolic, a property which is established in the present treatment by making use of a recent characterisation of the elementary spaces by Tits [1952, 1955]. The third important property is that space-time coordinates are related to homogeneous coordinates in a three-dimensional hyperbolic space. Consequently there is an isomorphism between homogeneous Lorentz transformations and transformations of homogeneous coordinates in hyperbolic space.

Our primary aim is to clarify the foundations of special relativity so that the theory becomes as acceptable and familiar as euclidean geometry. Accordingly, the question of independence of the axioms is of secondary importance and is briefly discussed in Chapter 10. Consistency of the axioms can be easily verified by considering the usual model of Minkowski space-time. Since much of the terminology and notation is new, a listing of definitions and notation has been included before the main text.

CHAPTER 2

KINEMATIC AXIOMS FOR MINKOWSKI SPACE-TIME

§2.1 Primitive Notions

A model of Minkowski space-time will be described in terms of the following primitive notions:

(i) a set $\mathcal{P}$ whose elements are called *particles*, each particle being a set whose elements are called *instants*; and

(ii) a binary *relation* σ defined on the set of all instants.

Particles are denoted by the symbols $\underset{\sim}{Q}$, $\underset{\sim}{R}$, $\underset{\sim}{S}$, $\underset{\sim}{T}$, $\underset{\sim}{U}$, $\underset{\sim}{V}$, $\underset{\sim}{W}$, ... Instants belonging to a particle are denoted by the particle symbol together with some subscript, for example, Q_1, Q_α, Q_a, $Q_x \in \underset{\sim}{Q}$. The set of all instants is denoted by $\mathcal{I}$. The binary relation σ is called the *signal relation*. An expression such as $Q_x \; \sigma \; R_y$ is to be read as *"a signal goes from* Q_x *to* R_y*"* or *"a signal leaves* Q_x *and arrives at* R_y*"*.

In this axiomatic system, the undefined basis of primitive notions is an adaptation of the basis used by Walker [1948]. Walker's undefined elements are *instants* and *particles*, and the undefined relations are the *signal correspondence relation* and the *temporal order relation*. In the present treatment, the signal relation is analogous to Walker's signal correspondence relation but the temporal order relation is defined in terms of the signal relation (§2.3). That is, *the present system makes use of only one undefined binary relation,* whereas the system of Walker is expressed in terms of *two* undefined binary relations. In physical terms, particles correspond to "inertial particles" and signals correspond to "light signals".

§2.2 Existence of Signal Functions

AXIOM I (SIGNAL AXIOM)

Given particles $\underset{\sim}{Q}, \underset{\sim}{R}$ *and an instant* $R_y \in \underset{\sim}{R}$,
there is a unique instant $Q_x \in \underset{\sim}{Q}$ *such that* $Q_x \, \sigma \, R_y$, *and*
there is a unique instant $Q_z \in \underset{\sim}{Q}$ *such that* $R_y \, \sigma \, Q_z$.

This axiom is used in the proof of Theorems 1 (§2.4), 2 (§2.6), 3 (§2.7), 13 (§3.6), 16 (§4.1), 17 (§4.2), 32 (§6.4) and 61 (§9.5).

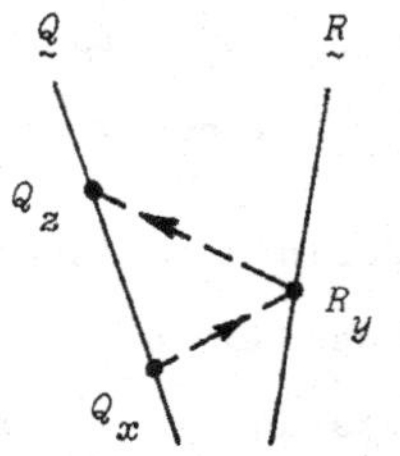

Fig. 1. In all diagrams, particles are represented by solid lines and signal relations are indicated by broken lines between instants which are represented by dots.

The Signal Axiom implies the existence of a bijection from $\underset{\sim}{Q}$ to $\underset{\sim}{R}$, which will be called a *signal function* and will be denoted by the symbol f_{RQ} where

$$f_{RQ}: \underset{\sim}{Q} \rightarrow \underset{\sim}{R}$$
$$Q_x \mapsto R_y \text{ if and only if } Q_x \; \sigma \; R_y.$$

Thus the Signal Axiom (Axiom I) is equivalent to the Signal Axiom of Walker [1948, Axiom S1, P322]. In the present treatment, the signal functions are equivalent to the "signal correspondences" and "signal mappings" of Walker [1948] and Walker [1959], respectively. The Axiom of Connectedness of Szekeres [1968, Axiom A4, P138] has a similar "physical content".

Given any two particles $\underset{\sim}{Q}$ and $\underset{\sim}{S}$, the composition of signal functions

$$f_{QS} \circ f_{SQ}$$

is a mapping from $\underset{\sim}{Q}$ to $\underset{\sim}{Q}$ which is related to the "motion of $\underset{\sim}{S}$ relative to $\underset{\sim}{Q}$" and is called the *record function* (of $\underset{\sim}{S}$ relative to $\underset{\sim}{Q}$).

§2.3 The Temporal Order Relation

In the present system, the "temporal (order) relation" is defined in terms of the σ-relation. This is a departure from the system of Walker [1948, P321], in which "temporal order" was an independent undefined relation. As a result of this difference in approach, the content of an axiom of Walker [1948, Axiom S3, P322] appears in the present treatment as a theorem (Theorem 1, §2.5).

Given a particle $\underset{\sim}{Q}$ and two instants Q_x, $Q_z \in \underset{\sim}{Q}$, if Q_x and Q_z are the same instant, we write $Q_x = Q_z$; otherwise we write $Q_x \neq Q_z$ and say that Q_x and Q_z are *distinct instants*. If $Q_x \neq Q_z$ and if there exists a particle $\underset{\sim}{R}$ with an instant $R_y \in \underset{\sim}{R}$ such that

$$Q_x \ \sigma \ R_y \quad \text{and} \quad R_y \ \sigma \ Q_z$$

we write $Q_x < Q_z$ and say that Q_x is *before* Q_z and Q_z is *after* Q_x. The relation $<$ is called the *(strict) temporal relation* (see the previous Fig. 1). The next axiom states that the strict temporal relation is antisymmetric; that is, it ensures that the previous definition is independent of the choice of $\underset{\sim}{R}$.

AXIOM II (FIRST AXIOM OF TEMPORAL ORDER)

Given a particle $\underset{\sim}{Q}$ *and two distinct instants* Q_x, $Q_z \in \underset{\sim}{Q}$, *then either* $Q_x < Q_z$, *or* $Q_z < Q_x$, *but not both.*

That is, the temporal relation is antisymmetric, and any two instants from the same particle are "comparable" in the sense that, given $Q_x, Q_z \in \underset{\sim}{Q}$, there are three mutually exclusive possibilities: (i) $Q_x = Q_z$ or (ii) $Q_x < Q_z$ or

(iii) $Q_z < Q_x$. This axiom is used in the proof of Theorems 1 (§2.5) and 11 (§3.4).

The symbol $\leqslant$ will be used so that statements of the form:

$$"Q_x = Q_z \text{ or } Q_x < Q_z",$$

can be written concisely as

$$"Q_x \leqslant Q_z".$$

We define the symbols $>$ and $\geqslant$ so that the statement "$Q_x > Q_z$" has the same meaning as the statement "$Q_z < Q_x$", and the statement "$Q_x \geqslant Q_z$" has the same meaning as the statement "$Q_z \leqslant Q_x$".

AXIOM III (SECOND AXIOM OF TEMPORAL ORDER)

Given a particle $\underset{\sim}{Q}$ *and instants* $Q_x, Q_y, Q_z \in \underset{\sim}{Q}$; *if* $Q_x < Q_y$ *and* $Q_y < Q_z$, *then* $Q_x < Q_z$.

That is, the temporal relation is transitive. Furthermore, the conclusion of the axiom implies that there exists a particle $\underset{\sim}{R}$ and an instant $R_y \in \underset{\sim}{R}$ such that

$$Q_x \,\sigma\, R_y \text{ and } R_y \,\sigma\, Q_z.$$

This axiom is used in Theorems 1 (§2.5) and 10 (§3.3).

An immediate corollary to this axiom is the proposition:

$$Q_x \leqslant Q_y \text{ and } Q_y \leqslant Q_z \Rightarrow Q_x \leqslant Q_z.$$

The previous two axioms imply that *the temporal order relation is a simple ordering on each particle, so we shall also call it the temporal order relation.*

§2.4 The Triangle Inequality

Composite statements of the form $Q_x \sigma R_y$ and $R_y \sigma S_z$ are sometimes combined for the sake of brevity to $Q_x \sigma R_y \sigma S_z$. (Note that σ is not a transitive relation and so $Q_x \sigma R_y \sigma S_z \not\Rightarrow Q_x \sigma S_z$). A statement containing one σ-relation is called a *direct signal*, and a composite statement involving two or more σ-relations is called an *indirect signal* (for example, $Q_x \sigma R_y$ is a direct signal and $Q_x \sigma R_y \sigma S_z$ is an indirect signal).

AXIOM IV (TRIANGLE INEQUALITY)

Let $\underset{\sim}{Q}, \underset{\sim}{R}, \underset{\sim}{S}$ *be particles with instants* $Q_w \in \underset{\sim}{Q}$, $R_y \in \underset{\sim}{R}$ *and* $S_x, S_z \in \underset{\sim}{S}$.
If $Q_w \sigma S_x$ *and* $Q_w \sigma R_y \sigma S_z$, *then* $S_x \leqslant S_z$.

That is, "of all signals leaving Q_x and arriving at $\underset{\sim}{S}$, no signal arrives before the direct signal". This axiom is equivalent to an axiom of Walker [1948, Axiom S2, P322]. Axiom IV is used in the proof of Theorems 1 (§2.5), 2 (§2.6), 3 (§2.7), 5 (§2.9), 6 (§2.9), 8 (§3.2), 28 (§6.1), and 57 (§9.1).

§2.5 Signal Functions are Order-Preserving

THEOREM 1 (Signal Functions are Order-Preserving)

Let $\underset{\sim}{S},\underset{\sim}{T}$ *be particles with instants* $S_x, S_z \in \underset{\sim}{S}$ *and instants* $T_x, T_z \in \underset{\sim}{T}$ *such that*

$$S_x \ \sigma \ T_x \text{ and } S_z \ \sigma \ T_z \ .$$

If $S_x < S_z$ *then* $T_x < T_z$.

This theorem is a consequence of Axioms I (§2.2), II (§2.3), III (§2.3) and IV (§2.4); and is used in the proof of Theorems 3 (§2.7), 5 (§2.9), 8 (§3.2), 17 (§4.2) and 60 (§9.4).

PROOF (see Fig. 2).

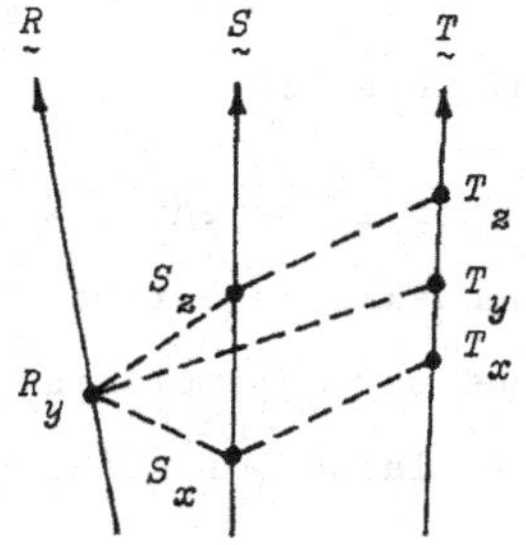

Fig 2

By the First Axiom of Temporal Order (Axiom II, §2.3) there is a particle $\underset{\sim}{R}$ with an instant $R_y \in \underset{\sim}{R}$ such that $S_x \,\sigma\, R_y \,\sigma\, S_z$. By the Signal Axiom (Axiom I, §2.2) there is an instant $T_y \in \underset{\sim}{T}$ such that $R_y \,\sigma\, T_y$. By the Triangle Inequality (Axiom IV, §2.4),

$$S_x \,\sigma\, T_x \text{ and } S_x \,\sigma\, R_y \,\sigma\, T_y \Rightarrow T_x \leqslant T_y,$$

$$R_y \,\sigma\, T_y \text{ and } R_y \,\sigma\, S_z \,\sigma\, T_z \Rightarrow T_y \leqslant T_z,$$

and by the Second Axiom of Temporal Order (Axiom III, §2.3),

$$T_x \leqslant T_y \text{ and } T_y \leqslant T_z \Rightarrow T_x \leqslant T_z.$$

By the Signal Axiom (Axiom I, §2.2), $T_x \neq T_z$, so $T_x < T_z$. □

The content of this theorem appeared in Walker's system [1948, Axiom S3, P322] as an axiom.

§2.6 The Coincidence Relation. Events.

Before defining the notion of an "event", which is the usual undefined element of Minkowski space-time, we define a relation of çoincidence on the set of instants. Events are then defined as equivalence classes of coincident instants.

Given instants Q_x and R_y such that both

$$Q_x \,\sigma\, R_y \text{ and } R_y \,\sigma\, Q_x,$$

we say that Q_x and R_y are *coincident instants* and write $Q_x \simeq R_y$.

THEOREM 2

The coincidence relation is symmetric and transitive.

This theorem is a consequence of Axiom I (§2.2) and Axiom IV (§2.4). It is applied in the definition of an event (this section and also §2.12) and in the proof of the Corollary to Theorem 7 (§2.12).

PROOF. By definition the coincidence relation is symmetric. By the description of our model, each instant belongs to some particle, so we consider any three particles $\underset{\sim}{Q}, \underset{\sim}{R}, \underset{\sim}{S}$ and any three instants $Q_x \in \underset{\sim}{Q}$, $R_y \in \underset{\sim}{R}$, $S_z \in \underset{\sim}{S}$ such that

$$Q_x \simeq R_y \text{ and } R_y \simeq S_z.$$

We will show that $Q_x \simeq S_z$. By the Signal Axiom (Axiom I, §2.2) there is an instant $S_u \in \underset{\sim}{S}$ such that $Q_x \, \sigma \, S_u$ and since $Q_x \, \sigma \, R_y \, \sigma \, S_z$, the Triangle Inequality (Axiom IV, §2.4) implies that $S_u \leq S_z$. Similarly $R_y \, \sigma \, S_z$ and $R_y \, \sigma \, Q_x \, \sigma \, S_u$ imply that $S_z \leq S_u$. Consequently $S_z = S_u$ and therefore $Q_x \, \sigma \, S_z$. A similar argument shows that $S_z \, \sigma \, Q_x$. □

At this stage we can not prove that the signal relation is reflexive. However the symmetry and transitivity of the coincidence relation allows us to deduce the following sufficient condition:

COROLLARY. *Given a particle* $\underset{\sim}{Q}$ *and an instant* $Q_x \varepsilon \underset{\sim}{Q}$, *if there is a particle* $\underset{\sim}{R}$ *and an instant* $R_y \varepsilon \underset{\sim}{R}$ *such that* $Q_x \simeq R_y$, *then* $Q_x \sigma Q_x$ *and hence* $Q_x \simeq Q_x$.

This corollary is used in the proof of Theorem 7 (§2.12) and its corollary where we show that the signal relation and the coincidence relation are reflexive.

PROOF. By transitivity. □

Now, given any instant Q_x, we call the equivalence class of coincident instants

$$[Q_x] \stackrel{def}{=} \{Q_x\} \cup \{R_y: \; R_y \simeq Q_x, \; R_y \varepsilon \underset{\sim}{R}, \; \underset{\sim}{R} \varepsilon \mathcal{P} \}$$

an *event* (see also §2.12). The set of all events is called *event space* and is denoted by the symbol $\mathcal{E}$.

Given particles $\underset{\sim}{Q},\underset{\sim}{R}$ and instants $Q_x \varepsilon \underset{\sim}{Q}$, $R_y \varepsilon \underset{\sim}{R}$ such that $Q_x \simeq R_y$, we say that the *particles* $\underset{\sim}{Q},\underset{\sim}{R}$ *coincide at the event* $[Q_x]$, which is equivalent to saying that the particles $\underset{\sim}{Q},\underset{\sim}{R}$ coincide at the event $[R_y]$. The coincidence relation has been defined and discussed previously by Walker [1948, P323], although he did not define events as equivalence classes of coincident instants.

§2.7 Optical Lines

Given instants Q_x, R_y, S_z such that

$$Q_x \; \sigma \; R_y \; \sigma \; S_z \text{ and } Q_x \; \sigma \; S_z$$

we say that the set of instants $\{Q_x, R_y, S_z\}$ is *in optical line*, which we denote by $|Q_x, R_y, S_z>$. Similarly, we say that the set of n instants $\{Q^{(1)}_1, Q^2_2, \ldots Q^n_{(n)}\}$ is *in optical line* if and only if, for all a, b, c with $1 \leqslant a \leqslant b \leqslant c \leqslant n$,

$$|Q^{(a)}_a, Q^{(b)}_b, Q^c_{(c)}> \; .$$

Note that this relation is *not* an order relation. Sometimes, instead of saying that Q_x, R_y, S_z are in optical line, we may use the alternative expression: S_z is in optical line with Q_x and R_y. This concept is intended to be analogous to the geometrical concept of collinearity.

THEOREM 3. *Let $\underset{\sim}{Q}, \underset{\sim}{R}, \underset{\sim}{S}, \underset{\sim}{T}, \underset{\sim}{U}$ be particles with instants $Q_1 \in \underset{\sim}{Q}$, $R_2 \in \underset{\sim}{R}$, $S_3 \in \underset{\sim}{S}$, $T_4 \in \underset{\sim}{T}$, and $U_5 \in \underset{\sim}{U}$. Then*

(i) $|Q_1, R_2, T_4\rangle$ *and* $|R_2, S_3, T_4\rangle$ *imply* $|Q_1, R_2, S_3, T_4\rangle$, *and*

(ii) $|R_2, S_3, T_4\rangle$ *and* $|R_2, T_4, U_5\rangle$ *imply* $|R_2, S_3, T_4, U_5\rangle$.

This theorem is a consequence of Axioms I (§2.2) and IV (§2.4) and Theorem 1 (§2.5); and is used in the proof of Theorems 4 (§2.7), 16 (§4.1) and 20 (§4.4).

PROOF. (i) By data,

(1) $$Q_1 \;\sigma\; R_2 \;\sigma\; T_4 \text{ and } Q_1 \;\sigma\; T_4, \text{ and}$$

(2) $$R_2 \;\sigma\; S_3 \;\sigma\; T_4 \text{ and } R_2 \;\sigma\; T_4.$$

By the Signal Axiom (Axiom I, §2.2) and the Triangle Inequality (Axiom IV, §2.4), there is an instant $S_x \in \underset{\sim}{S}$ and an instant $T_x \in \underset{\sim}{T}$, with $S_x \leqslant S_3$, such that

$$Q_1 \;\sigma\; S_x \;\sigma\; T_x$$

and since signal functions are order-preserving (Theorem 1, §2.5), $T_x \leqslant T_4$. But $Q_1 \;\sigma\; T_4$ and so the Triangle Inequality (Axiom IV, §2.4) implies that $T_4 \leqslant T_x$, whence $T_x = T_4$. Therefore by the Signal Axiom (Axiom I, §2.2) and (2),

$$Q_1 \;\sigma\; S_3 \;\sigma\; T_4, \text{ and by (1)}$$

(3) $$|Q_1, S_3, T_4\rangle \;.$$

Also, from (1), (2) and (3),

(4) $$|Q_1, R_2, S_3> .$$

The data, together with (3) and (4) are equivalent to $|Q_1, R_2, S_3, T_4>$.

(ii) The proof of (ii) is similar. □

COROLLARY. *Let* $\underset{\sim}{Q}, \underset{\sim}{R}, \underset{\sim}{S}, \underset{\sim}{T}, \underset{\sim}{U}$ *be particles with instants* $Q_1 \in \underset{\sim}{Q}$, $R_2 \in \underset{\sim}{R}$, $S_3 \in \underset{\sim}{S}$, $T_4 \in \underset{\sim}{T}$, $U_5 \in \underset{\sim}{U}$.

Then (i) $|Q_1, R_2, S_3, U_5>$ *and* $|S_3, T_4, U_5> \Rightarrow |Q_1, R_2, S_3, T_4, U_5>$, *and*

(ii) $|Q_1, S_3, T_4, U_5>$ *and* $|Q_1, R_2, S_3> \Rightarrow |Q_1, R_2, S_3, T_4, U_5>$.

This corollary is used in the proof of Theorem 4 (§2.7).

PROOF. (i) $|R_2, S_3, U_5>$ and $|S_3, T_4, U_5> \Rightarrow |R_2, S_3, T_4, U_5>$

$|Q_1, S_3, U_5>$ and $|S_3, T_4, U_5> \Rightarrow |Q_1, S_3, T_4, U_5>$

$|Q_1, R_2, S_3>$ and $|Q_1, S_3, T_4> \Rightarrow |Q_1, R_2, S_3, T_4>$

From these relations and the data we have $|R_2, T_4, U_5>$, $|Q_1, T_4, U_5>$, $|Q_1, R_2, U_5>$, $|Q_1, R_2, T_4>$ which together are equivalent to $|Q_1, R_2, T_4, U_5>$. Now the data and the four relations between quadruples of instants are equivalent to $|Q_1, R_2, S_3, T_4, U_5>$.

(ii) The proof is similar. □

If Q_x, R_x, S_x are non-coincident instants such that either

$|Q_x, R_x, S_x>$ or $|S_x, Q_x, R_x>$ or $|R_x, Q_x, S_x>$ or $|S_x, R_x, Q_x>$,

we say that the instant S_x is *exterior to* the (pair of) instants Q_x and R_x.

AXIOM V (UNIQUENESS OF EXTENSION OF OPTICAL LINES)

Let Q_x and R_x be any two non-coincident instants such that $Q_x \sigma R_x$. If S_x and T_x are any two instants exterior to Q_x and R_x, then Q_x and R_x are in optical line with S_x and T_x.

This axiom is used in the proof of Theorems 4 (§2.7), 16 (§4.1) and 20 (§4.4).

An equivalent, though apparently weaker, statement is: "If S_x and T_x are any two instants exterior to Q_x and R_x, then at least one of Q_x and R_x is in optical line with S_x and T_x". The demonstration of logical equivalence involves a separate simple procf for each possible arrangement of instants. The Axiom of Uniqueness of Extension of Optical Lines is analogous to the axiom of Uniqueness of Prolongation of Busemann [1955, §8.1] and is weaker than the corresponding axiom of Walker [1948, Axiom S.4., P324].

Thus the axioms which have been stated so far do not allow us to conclude, as in the treatment of Walker [1948, Theorem 6.1, P324], that: "All instants collinear optically with two non-coincident instants are in one optical line", since at this stage it is conceivable that we could have a situation analogous to multiple geodesics between antipodal points on a sphere. However we can prove the weaker theorem:

THEOREM 4. (Existence of an Optical Line)

Given particles $\underset{\sim}{Q}, \underset{\sim}{R}, \underset{\sim}{S}$ *with instants* $Q_1 \in \underset{\sim}{Q}$, $R_2 \in \underset{\sim}{R}$, $S_3 \in \underset{\sim}{S}$ *such that*

$$|Q_1, R_2, S_3> \textit{ and } Q_1 \neq R_2 \neq S_3,$$

then all instants which are in optical line with Q_1 *and* R_2, *or with* R_2 *and* S_3, *are in optical line.*

A maximal set of instants, all of which are in optical line, is called *an optical line*. Thus, this theorem states that *an optical line is uniquely determined by any three distinct instants which are in optical line.*

This theorem is a consequence of Axiom V (§2.7) and Theorem 3 (§2.7), together with its corollary; and is used in the proof of Theorems 22 (§5.2), 26 (§5.5), 27 (§6.1), Corollary 1 to Theorem 30 (§6.3), and Theorem 32 (§6.4).

PROOF. We consider any two instants T_x, U_y such that either

(i) $|T_x, Q_1, R_2>$ or (ii) $|Q_1, T_x, R_2>$ or (iii) $|Q_1, R_2, T_x>$ or

(iv) $|T_x, R_2, S_3>$ or (v) $|R_2, T_x, S_3>$ or (vi) $|R_2, S_3, T_x>$,

and either

(1) $|U_y, Q_1, R_2>$ or (2) $|Q_1, U_y, R_2>$ or (3) $|Q_1, R_2, U_y>$ or

(4) $|U_y, R_2, S_3>$ or (5) $|R_2, U_y, S_3>$ or (6) $|R_2, S_3, U_y>$.

We must show that for any instants T_x and U_y satisfying (i) - (vi) and (1) - (6), respectively, the instants Q_1, R_2, S_3, T_x, U_y are in optical line in some order. By data we know that $|Q_1, R_2, S_3>$ and we apply the Axiom of Uniqueness of Extension of Optical Lines (Axiom V, §2.7) and Theorem 3 (§2.7) to obtain the following implications:

(i) $\Rightarrow |T_x, Q_1, R_2, S_3>$ (by Axiom V),
(ii) $\Rightarrow |Q_1, T_x, R_2, S_3>$ (by Theorem 3),
(iii) $\Rightarrow |Q_1, R_2, T_x, S_3>$ or $|Q_1, R_2, S_3, T_x>$ (by Axiom V),
(iv) $\Rightarrow |Q_1, T_x, R_2, S_3>$ or $|T_x, Q_1, R_2, S_3>$ (by Axiom V),
(v) $\Rightarrow |Q_1, R_2, T_x, S_3>$ (by Theorem 3),
(vi) $\Rightarrow |Q_1, R_2, S_3, T_x>$ (by Axiom V),

and

(1) $\Rightarrow |U_y, Q_1, R_2, S_3>$ (by Axiom V),
(2) $\Rightarrow |Q_1, U_y, R_2, S_3>$ (by Theorem 3),
(3) $\Rightarrow |Q_1, R_2, U_y, S_3>$ or $|Q_1, R_2, S_3, U_y>$ (by Axiom V),
(4) $\Rightarrow |Q_1, U_y, R_2, S_3>$ or $|U_y, Q_1, R_2, S_3>$ (by Axiom V),
(5) $\Rightarrow |Q_1, R_2, U_y, S_3>$ (by Theorem 3), and
(6) $\Rightarrow |Q_1, R_2, S_3, U_y>$ (by Axiom V).

Now each of the 36 cases (i)(1)$\cdots$(i)(6),$\cdots$,(vi)(1)$\cdots$(vi)(6) can be considered separately by applying Axiom V and the corollary to Theorem 3; however, it should be apparent that similar considerations will apply to cases *(n)(M)* and *(6-n)(6-M)*, so it is sufficient to consider the first 18 cases. Of these first 18 cases, all can be proved using Axiom V, except for the following cases in which the corollary to Theorem 3 is also applied:

Case (ii)(3) $|Q_1,T_x,R_2,S_3>$ and $(|Q_1,R_2,U_y,S_3>$ or $|Q_1,R_2,S_3,U_y>)$ $\Rightarrow |Q_1,T_x,R_2,U_y,S_3>$ (by the corollary to Theorem 3) or $|Q_1,T_x,R_2,S_3,U_y>$ (by Axiom V).

Case (ii)(5) $|Q_1,T_x,R_2,S_3>$ and $|Q_1,R_2,U_y,S_3> \Rightarrow |Q_1,T_x,R_2,U_y,S_3>$.

Case (iii)(2) $(|Q_1,R_2,T_x,S_3>$ or $|Q_1,R_2,S_3,T_x>)$ and $|Q_1,U_y,R_2,S_3> \Rightarrow |Q_1,U_y,R_2,T_x,S_3>$ (by the corollary to Theorem 3) or $|Q_1,U_y,R_2,S_3,T_x>$ (by the corollary to Theorem 3).

Case (iii)(4) $(|Q_1,R_2,T_x,S_3>$ or $|Q_1,R_2,S_3,T_x>)$ and $(|Q_1,U_y,R_2,S_3)$ or $|U_y,Q_1,R_2,S_3>) \Rightarrow |Q_1,U_y,R_2,T_x,S_3>$ or $|U_y,Q_1,R_2,T_x,S_3>$ or $|Q_1,U_y,R_2,S_3,T_x>$ (by the corollary to Theorem 3), or $|U_y,Q_1,R_2,S_3,T_x>$ (by Axiom V).

The remaining first 14 cases are all simple applications of Axiom V, and are not included here for the sake of brevity. □

This theorem does not exclude the possibility of instants which are not exterior to Q_1 and S_3 and which may not be in optical line with the pair of instants Q_1 and R_2, or with the pair R_2 and S_3. A stronger result is obtained in Corollary 1 of Theorem 33 (§6.4), where we show that an optical line is uniquely determined by any two non-coincident signal-related instants.

§2.8 Axiom of the Intermediate Particle

Two particles $\underset{\sim}{Q}, \underset{\sim}{R}$ are *permanently coincident* if, for each $Q_x \in \underset{\sim}{Q}$, there is some $R_y \in \underset{\sim}{R}$ such that $Q_x = R_y$. We denote permanent coincidence of $\underset{\sim}{Q}$ and $\underset{\sim}{R}$ by writing $\underset{\sim}{Q} \simeq \underset{\sim}{R}$. By Theorem 2 (§2.6) permanent coincidence of particles is a symmetric and transitive relation. We say that particles $\underset{\sim}{Q}, \underset{\sim}{R}$ are *distinct* if $\underset{\sim}{Q} \not\simeq \underset{\sim}{R}$.

Given particles $\underset{\sim}{Q}, \underset{\sim}{R}, \underset{\sim}{S}$ and an instant $R_x \in \underset{\sim}{R}$ such that

$$|f^{-1}_{RQ}(R_x), R_x, f_{SR}(R_x)\rangle \text{ and } |f^{-1}_{RS}(R_x), R_x, f_{QR}(R_x)\rangle,$$

we say that *the instants* R_x *is between the particles* $\underset{\sim}{Q}$ *and* $\underset{\sim}{S}$.

If, for all $R_x \in \underset{\sim}{R}$, the instant R_x is between the particles $\underset{\sim}{Q}$ and $\underset{\sim}{S}$, we say that *the particle* $\underset{\sim}{R}$ *is between the particles* $\underset{\sim}{Q}$ *and* $\underset{\sim}{S}$*;* and we denote this by writing $\langle \underset{\sim}{Q}, \underset{\sim}{R}, \underset{\sim}{S} \rangle$.

AXIOM VI (INTERMEDIATE PARTICLE)

Given distinct particles $\underset{\sim}{Q}, \underset{\sim}{S}$ *and instants* $Q_c \in \underset{\sim}{Q}$, $S_c \in \underset{\sim}{S}$ *such that* $Q_c \approx S_c$, *there exists a particle* $\underset{\sim}{R}$ *such that*

$$\langle \underset{\sim}{Q}, \underset{\sim}{R}, \underset{\sim}{S} \rangle \text{ and } \underset{\sim}{Q} \neq \underset{\sim}{R} \neq \underset{\sim}{S}.$$

That is, there is a particle between $\underset{\sim}{Q}$ and $\underset{\sim}{S}$, which is distinct from both. This axiom is used in the proof of Theorems 22 (§5.2) and 28 (§6.1).

§2.9 The Isotropy of SPRAYs

Any set of particles which coincide simultaneously at a given event is called a *SPRAY*. We define

$$SPR[Q_c] \overset{def}{=} \{\underset{\sim}{R}: f_{QR} \circ f_{RQ}(Q_c) = Q_c, \underset{\sim}{R} \in \mathcal{P}\}.$$

That is, *SPR*$[Q_c]$ *is the set of particles which coincide (with* $\underset{\sim}{Q}$*) at the event* $[Q_c]$ (see Fig. 3). A subset of a SPRAY is called a *sub-SPRAY*.

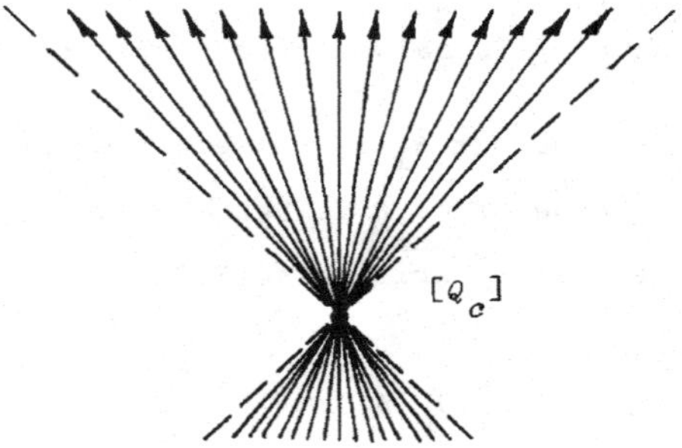

Fig. 3. In Minkowski space-time, $SPR[Q_c]$ is the set of "inertial particles whose paths are contained within the light cone whose vertex is the event $[Q_c]$". In this and subsequent diagrams, events are represented by dots.

The set of instants belonging to the particles of a SPRAY is called a *spray*. We define

$$spr[Q_c] = \{R_x : R_x \in \underset{\sim}{R},\ \underset{\sim}{R} \in SPR[Q_c]\} .$$

A spray restricted to a sub-SPRAY is called a *sub-spray*. The next axiom states that all SPRAYs are isotropic. In the present treatment, it is this axiom which expresses the "Principle of Relativity" of Einstein [1905, §2]. In the theory of euclidean geometry, a stronger analogue of this axiom has been used by Pogorolev [1966, Axiom III_7, Ch. II, §3] who called his axiom an "axiom of motion".

AXIOM VII (ISOTROPY OF SPRAYS)

Let $\underset{\sim}{Q},\underset{\sim}{R},\underset{\sim}{S}$ *be distinct particles with instants* $Q_c \in \underset{\sim}{Q}$, $R_c \in \underset{\sim}{R}$, $S_c \in \underset{\sim}{S}$ *such that* $Q_c \simeq R_c \simeq S_c$.
If, for some instant $Q_x \in \underset{\sim}{Q}$ *with* $Q_x \neq Q_c$,

$$f_{QR} \circ f_{RQ}(Q_x) = f_{QS} \circ f_{SQ}(Q_x),$$

then there is an injection ϕ *from* $spr[Q_c]$ *to* $spr[Q_c]$ *such that:*

(i) $\underset{\sim}{T} \in SPR[Q_c] \Longrightarrow \phi(\underset{\sim}{T}) \in SPR[Q_c]$,

(ii) *for all particles* $\underset{\sim}{T},\underset{\sim}{U} \in SPR[Q_c]$,

$$f_{TU} \circ f_{UT}(T_x) = T_z \Longrightarrow f_{\phi(T)\phi(U)} \circ f_{\phi(U)\phi(T)}(\phi(T_x)) = \phi(T_z),$$

(iii) *for all* $Q_x \in \underset{\sim}{Q}$, $\phi(Q_x) \simeq Q_x$, *and*

(iv) $\phi(\underset{\sim}{R}) \simeq \underset{\sim}{S}$.

It follows immediately that:

$$\phi(\underset{\sim}{Q}) \simeq \underset{\sim}{Q}.$$

This axiom is used in the proof of Theorems 5 (§2.9), 6(§2.9), 24(§5.3), 42(§7.3) and 57 (§9.1).

The mapping ϕ is called an *isotropy mapping*. The previous statements mean that:-

(i) ϕ maps particles onto particles;

(ii) ϕ is a homomorphism. A stronger property than (ii), which is more obviously a homomorphism, is the following:

(ii') "For all particles $\underset{\sim}{T}, \underset{\sim}{U} \in SPR[Q_c]$ and for any instants $T_x \in \underset{\sim}{T}, U_y \in \underset{\sim}{U}$,

$$T_x \; \sigma \; U_y \Longrightarrow \phi(T_x) \; \sigma \; \phi(T_y)" ,$$

however in the present axiomatic system it is sufficient to assume (ii);

(iii) each instant of $\underset{\sim}{Q}$ is mapped onto an instant coincident with itself. This is a weaker statement than:

(iii') "each instant of $\underset{\sim}{Q}$ is invariant", which is not assumed in the present axiomatic system;

(iv) $\underset{\sim}{R}$ is mapped onto a particle which is permanently coincident with $\underset{\sim}{S}$. This is a weaker statement than:

(iv') "$\underset{\sim}{R}$ is mapped onto $\underset{\sim}{S}$", which also is not assumed in the present axiomatic system; and finally

the statement following the axiom means that: $\underset{\sim}{Q}$ is mapped onto a particle which is permanently coincident with $\underset{\sim}{Q}$, which is a weaker statement than:

"$\underset{\sim}{Q}$ is invariant", which can not be proved in this system.

It may be worth noting that statements (ii'), (iii'), (iv') likewise can not be proved in the present axiomatic system, since many particles can be permanently coincident or "indistinct" as "observed by other particles". This is a consequence of choosing instants, rather than events, as the fundamental undefined elements.

THEOREM 5. *Let $\underset{\sim}{Q},\underset{\sim}{R},\underset{\sim}{S}$ be particles in $SPR[Q_c]$, as in the preceding axiom, and let $\underset{\sim}{T}$ be any particle in $SPR[Q_c]$. Then*

(i) $f_{QR} \circ f_{RQ} = f_{QS} \circ f_{SQ}$, *and*

(ii) $f_{QT} \circ f_{TQ} = f_{Q\phi(T)} \circ f_{\phi(T)Q}$.

This theorem is a consequence of Axioms IV (§2.4) and VII (§2.9) and Theorem 1 (§2.5). It is used in the proof of Theorem 6 (§2.9), Corollary 2 of Theorem 22 (§5.2) and Theorems 23 (§5.3) and 30 (§6.3).

PROOF. (i) By the Triangle Inequality (Axiom IV, §2.4)

$$f_{Q\phi(R)} \circ f_{\phi(R)Q} \leq f_{QS} \circ f_{S\phi(R)} \circ f_{\phi(R)S} \circ f_{SQ}$$

But by the preceding axiom, $\phi(\underset{\sim}{R}) \simeq \underset{\sim}{S}$, so $f_{S\phi(R)} \circ f_{\phi(R)S}$ is an identity mapping and therefore

$$f_{Q\phi(R)} \circ f_{\phi(R)Q} \leq f_{QS} \circ f_{SQ}.$$

The opposite inequality is proved in a similar manner.

(ii) For each instant $Q_x \in \underset{\sim}{Q}$ there is an instant $Q_z \in \underset{\sim}{Q}$ such that

(1) $$f_{QT} \circ f_{TQ}(Q_x) = Q_z ,$$

and by part (ii) of the preceding axiom,

(2) $$f_{\phi(Q)\phi(T)} \circ f_{\phi(T)\phi(Q)}(\phi(Q_x)) = \phi(Q_z).$$

Also by part (iii) of the preceding axiom and the Triangle Inequality (Axiom IV, §2.4),

(3) $$\begin{aligned} f_{Q\phi(T)} \circ f_{\phi(T)Q}(Q_x) &\leqslant f_{Q\phi(Q)} \circ f_{\phi(Q)\phi(T)} \circ f_{\phi(T)\phi(Q)} \circ f_{\phi(Q)Q}(Q_x) \\ &= f_{Q\phi(Q)} \circ f_{\phi(Q)\phi(T)} \circ f_{\phi(T)\phi(Q)}(\phi(Q_x)) \\ &= f_{Q\phi(Q)}(\phi(Q_z)) \\ &= Q_z . \end{aligned}$$

Now if

(4) $$f_{Q\phi(T)} \circ f_{\phi(T)Q}(Q_x) = Q_y < Q_z,$$

then, as above,

(5) $$f_{\phi(Q)\phi(T)} \circ f_{\phi(T)\phi(Q)}(\phi(Q_x)) \leqslant$$

$$\leqslant f_{\phi(Q)Q} \circ f_{Q\phi(T)} \circ f_{\phi(T)Q} \circ f_{Q\phi(Q)}(\phi(Q_x))$$

$$= f_{\phi(Q)Q} \circ f_{Q\phi(T)} \circ f_{\phi(T)Q}(Q_x)$$

$$= f_{\phi(Q)Q}(Q_y) = \phi(Q_y)$$

$$< \phi(Q_z), \text{ by Theorem 1 (§2.5),}$$

which is a contradiction of (2). That is, (4) should be replaced by

(6) $$f_{Q\phi(T)} \circ f_{\phi(T)Q}(Q_x) = Q_z,$$

which, together with (1), completes the proof of (ii). □

The next result, which is a consequence of the previous two axioms, is equivalent to an axiom of Szekeres [1968, Axiom A.3, P137].

THEOREM 6 (Permanently Coincident Particles)

Two distinct particles coincide at no more than one event.

This theorem is a consequence of Axioms IV (§2.4) and VII (§2.9) and Theorem 5 (§2.9). It is used in the proof of Theorems 8 (§3.2), 11 (§3.4), 21 (§5.1), 25 (§5.4), 29 (§6.2), Corollary 1 of 30 (§6.3), 33 (§6.4), 34 (§6.4), 36 (§7.1) and 46 (§7.5).

PROOF. Given particles $\underset{\sim}{Q},\underset{\sim}{S}$ and instants $Q_0, Q_1 \in \underset{\sim}{Q}$ and $S_0, S_1 \in \underset{\sim}{S}$ such that

$$Q_0 \simeq S_0 \neq S_1 \simeq Q_1,$$

we will show that $\underset{\sim}{Q} \simeq \underset{\sim}{S}$.

Suppose the contrary; that is, suppose that $\underset{\sim}{Q} \not\simeq \underset{\sim}{S}$. By the Axiom of the Intermediate Particle (Axiom VI, §2.8), there is a particle $\underset{\sim}{R}$ such that

(1) $$\langle \underset{\sim}{Q},\underset{\sim}{R},\underset{\sim}{S} \rangle \text{ and } \underset{\sim}{Q} \not\simeq \underset{\sim}{R} \not\simeq \underset{\sim}{S}.$$

Since $Q_0 \simeq S_0, Q_1 \simeq S_1$ and $\langle \underset{\sim}{Q},\underset{\sim}{R},\underset{\sim}{S} \rangle$ it follows from the definition of "betweenness" that there are instants $R_0, R_1 \in \underset{\sim}{R}$ such that

$$Q_0 \simeq R_0 \simeq S_0 \text{ and } Q_1 \simeq R_1 \simeq S_1,$$

or equivalently,

$$f_{QR} \circ f_{RQ}(Q_1) = f_{QS} \circ f_{SQ}(Q_1) = Q_1.$$

By the Axiom of Isotropy of SPRAYs (Axiom VII, §2.9), but with Q_0 taking the place of Q_c, there is an isotropy mapping ϕ such that $\phi(\underset{\sim}{R}) \simeq \underset{\sim}{S}$, and by the previous theorem,

(2) $$f_{QR} \circ f_{RQ} = f_{QS} \circ f_{SQ}.$$

But statement (1) and the definition of "betweenness" (§2.8) imply that, for at least one instant $Q_x \in \underset{\sim}{Q}$,

(3) $$f_{QR} \circ f_{RQ}(Q_x) < f_{QS} \circ f_{SQ}(Q_x),$$

which contradicts (2). Thus the supposition that $\underset{\sim}{Q} \neq \underset{\sim}{S}$ is false, which completes the proof. □

§2.10 The Axiom of Dimension

In the theory of absolute geometry, we can specify the dimension n by the axiom: "there is a set of $n+1$ equidistant points, but there is no set of $n+2$ equidistant points". In the case of 3-dimensional absolute geometry, the set of four equidistant points is, of course, a tetrahedron. An axiom of this form implies that the set of points is non-empty; that is, an axiom of dimension is also an axiom of existence (for $n \geqslant 0$).

In the present treatment we do not have a measure of "distance"; however, we can compare the relative motions of particles from the same SPRAY by comparing their record functions. This leads to the following definition which is analogous to the concept of a set of equidistant points in a space with a non-symmetric metric.

Given an event $[Q_c]$, the sub-SPRAY of n particles, $\{\underset{\sim}{R}^{(i)}: i=1,\cdots,n;\ \underset{\sim}{R}^{(i)} \in SPR[Q_c]\}$ *is a symmetric sub-SPRAY.* if, for all $i \in \{1,\cdots,n\}$, there exists $R_x^{(i)} \in \underset{\sim}{R}^{(i)}$ with $R_x^{(i)} \notin [Q_c]$ such that, for all $j,k \in \{1,\cdots,n\}$, $j \neq i$, $k \neq i$,

$$f_{ij} \circ f_{ji}\left(R_x^{(i)}\right) = f_{ik} \circ f_{ki}\left(R_x^{(i)}\right),$$

where f_{ji} is the signal function which sends $\underset{\sim}{R}^{(i)}$ onto $\underset{\sim}{R}^{(j)}$. By Theorem 5 (§2.9) it follows that *a symmetric sub-SPRAY has the stronger property: for all* $i,j,k \in \{1,\cdots,n\}$, $j \neq i$, $k \neq i$,

$$f_{ij} \circ f_{ji} = f_{ik} \circ f_{ki} .$$

AXIOM VIII (DIMENSION)

There is a SPRAY which has a maximal symmetric sub-SPRAY of four distinct particles.

This axiom states that particles exist and ensures that the "velocity space" of each SPRAY is three-dimensional (see §9.1 and §9.5). The Axiom is used in the proof of Theorems 7 (§2.12), 11 (§3.4), 15 (§3.7), 25 (§5.4) and 61 (§9.5).

§2.11 The Axiom of Incidence

AXIOM IX (INCIDENCE)

Let $\underset{\sim}{Q}$ and $\underset{\sim}{R}$ be distinct particles with instants $Q_z \in \underset{\sim}{Q}$ and $R_z \in \underset{\sim}{R}$ such that $Q_z \simeq R_z$.
Given any instant $Q_x \in \underset{\sim}{Q}$ with $Q_x \neq Q_z$, there is some particle $\underset{\sim}{S}$ with instants $S_x, S_y \in \underset{\sim}{S}$ and an instant $R_y \in \underset{\sim}{R}$ with $R_y \neq R_z$, such that

$$Q_x \simeq S_x \text{ and } R_y \simeq S_y.$$

(see Fig. 4)

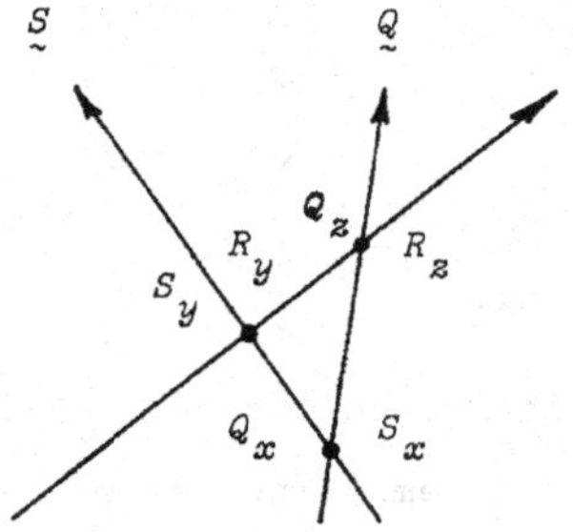

Fig. 4

Note that order relations have not been specified between the pairs of instants Q_x and Q_z, R_y and R_z, or S_x and S_y. The axiom is used in the proof of Theorems 7 (§2.12), 15 (§3.7), 25 (§5.4), 28 (§6.1) and 32 (§6.4).

§2.12 The Axiom of Connectedness

AXIOM X (CONNECTEDNESS)

The set of instants is connected in the following sense: given any two instants Q_x *and* T_z *there are particles* $\underset{\sim}{R}$ *and* $\underset{\sim}{S}$ *with instants* $R_x, R_y \in \underset{\sim}{R}$ *and* $S_y, S_z \in \underset{\sim}{S}$ *such that*

$$Q_x \simeq R_x \quad \text{and} \quad R_y \simeq S_y \quad \text{and} \quad S_z \simeq T_z$$

(see Fig. 5).

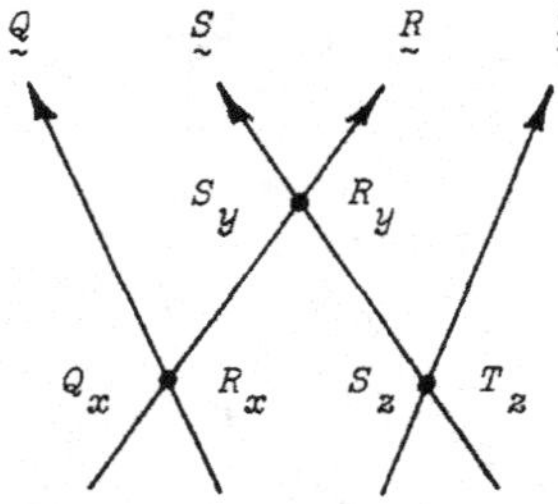

Fig. 5

This axiom has some resemblance to the first axiom of Euclid. It is independent of the remaining set of axioms, since a model can be constructed from a *4+1*-dimensional Minkowski space by considering two *3+1*-dimensional Minkowski sub-spaces having no event in common. The axiom is used in the proof of Theorems 7 (§2.12), 10 (§2.10), 15 (§3.7), 26 (§5.5), Corollary 2 of Theorem 33 (§6.4) and Theorem 61 (§9.5).

A physical interpretation of the next theorem is that "no particle can move as fast as a signal".

THEOREM 7. (The Signal Relation is Reflexive)

Given any particle $\underset{\sim}{Q}$ *and any instant* $Q_x \in \underset{\sim}{Q}$,

$$Q_x \ \sigma \ Q_x.$$

That is, f_{QQ} *is an identity mapping.*

This theorem is a consequence of Axioms VIII (§2.10), IX(§2.11), and X (§2.12) and the Corollary to Theorem 2 (§2.6). It is used in the proof of Theorem 16 (§4.1).

PROOF. In order to apply the corollary to Theorem 2 (§2.6) it is necessary to show that there is some particle $\underset{\sim}{S}$ with an instant $S_x \in \underset{\sim}{S}$ such that $S_x \simeq Q_x$. By the Axiom of Connectedness (Axiom X, §2.12) and the Axiom of Dimension (Axiom VIII, §2.10), there is a particle $\underset{\sim}{R}$ and instants $Q_c \in \underset{\sim}{Q}$ and $R_c \in \underset{\sim}{R}$ such that $Q_c \simeq R_c$. However we can not assert that $Q_c = Q_x$. For any instant $Q_x \neq Q_c$, the Axiom of Incidence (Axiom IX, §2.11) asserts that there is a particle $\underset{\sim}{S} \neq \underset{\sim}{Q}$ and an instant $S_x \in \underset{\sim}{S}$ such that $Q_x \simeq S_x$. Now the conditions of the Corollary to Theorem 2 (§2.6) are satisfied. □

COROLLARY.

(i) *The coincidence relation is an equivalence relation.*

(ii) *For any instant* Q_x, *the corresponding event* $[Q_x]$ *is given by*

$$[Q_x] = \{R_y : R_y \simeq Q_x,\ R_y \in \mathcal{I}\}\ .$$

PROOF. Part (i) is a consequence of Theorem 2 (§2.6). Part (ii) is an application of (i) to the definition of an event (§2.6). □

§2.13 Compactness of Bounded sub-SPRAYs.

Before stating the final axiom, which involves a property of compactness, it is necessary to define the concepts of "a bounded sub-SPRAY" and "a cluster particle". In several treatises on geometry an analogous property, with points taking the place of particles, has been taken as an axiom; the property has been called "finite compactness" by Busemann [1955, §2.6], and is a stronger assumption than the axiom of "continuity" of absolute geometry.

Given a particle $\underset{\sim}{Q}$, an instant $Q_c \in \underset{\sim}{Q}$, and a sub-SPRAY $\mathcal{A}[Q_c] \subseteq SPR[Q_c]$, we say that $\mathcal{A}[Q_c]$ is a *bounded sub-SPRAY* if, for some $\underset{\sim}{R} \in SPR[Q_c]$, there are instants $R_1, R_2 \in \underset{\sim}{R}$ with

$$f_{RQ}(Q_c) < R_1 < R_2 ,$$

such that

$$\mathcal{A}[Q_c] \subseteq \{\underset{\sim}{S}: \; f_{RS} \circ f_{SR}(R_1) \leq R_2, \; \underset{\sim}{S} \in SPR[Q_c]\}$$

(see Fig. 6).

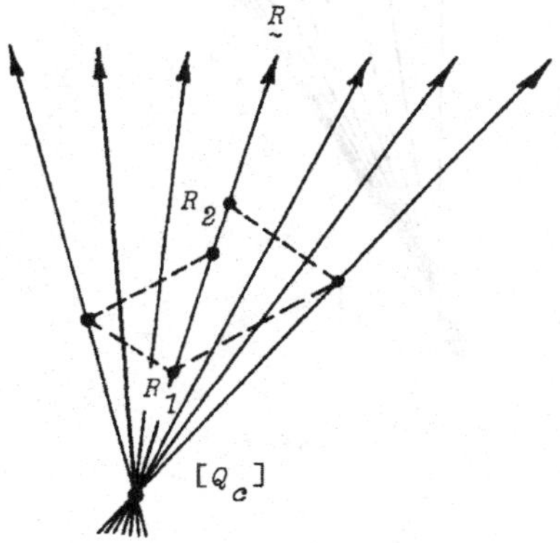

Fig. 6

Given a particle $\underset{\sim}{Q}$, an instant $Q_c \in \underset{\sim}{Q}$ and a sub-SPRAY $\mathcal{A}[Q_c] \subseteq SPR[Q_c]$, we say that $\underset{\sim}{S}$ is a *cluster particle* of $\mathcal{A}[Q_c]$, if (i) $\underset{\sim}{S} \in SPR[Q_c]$, and

(ii) for any instants $S_1, S_2 \in \underset{\sim}{S}$ with

$$f_{SQ}(Q_c) < S_1 < S_2 ,$$

there exists some $\underset{\sim}{T} \in \mathcal{A}[Q_c]$ such that

$$S_1 < f_{ST} \circ f_{TS}(S_1) < S_2$$

(see Fig. 7).

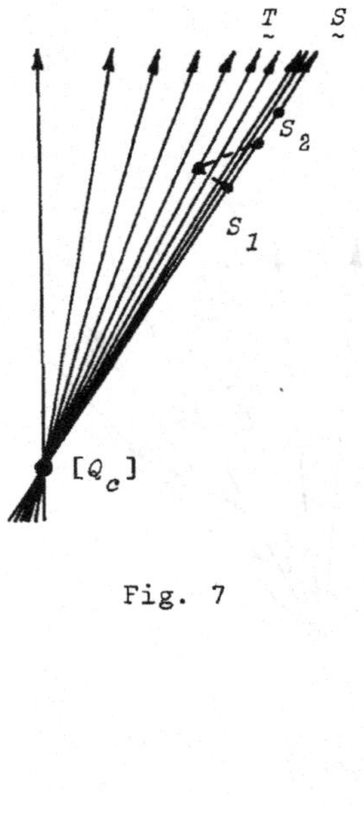

Fig. 7

By applying Theorem 1 (§2.5), condition (ii) could be replaced by the equivalent condition (ii') which is specified "relative to $\underset{\sim}{Q}$", and is:

(ii') given $Q_1, Q_2 \in \underset{\sim}{Q}$ with $Q_c < Q_1 < Q_2$, there exists some $\underset{\sim}{T} \in \mathcal{A}$, such that

$$Q_1 < f^{-1}_{SQ} \circ f_{ST} \circ f_{TS} \circ f_{SQ}(Q_1) < Q_2.$$

Note that the conditions

$$f_{SQ}(Q_c) < S_1 < S_2 \quad \text{or} \quad Q_c < Q_1 < Q_2$$

weaken the assumption of compactness so that it applies only "after coincidence". Without these (equivalent) restrictions, the following axiom would also apply "before coincidence": consequently the results of Chapter 5 would apply not only "after coincidence" but also "before coincidence" and Theorem 30 (§6.3) would be a simple corollary of Theorem 29 (§6.2).

AXIOM XI (COMPACTNESS)

Every bounded infinite sub-SPRAY has a cluster particle.

This is the last axiom. It is a stronger axiom than the "Axiom of Continuity" of absolute geometry, and yields similar results. It specifies properties of "velocity space", and also implies the "conditional completeness" of each particle regarded as a set of instants (see §5.4). It is used in the proof of Theorems 21 (§5.1) and 57 (§9.1).

CHAPTER 3

CONDITIONALLY COMPLETE PARTICLES

In this section, we embed each particle in its "conditional completion", so that every bounded monotone sequence of instants has a "limit" in the "conditionally completed particle". This simplifies later discussions and is justified by Theorem 25 (§5.4), which shows that all particles are "conditionally complete" as a consequence of the given axioms. A similar process of "completion" is employed in the arguments of Walker [1948, 1959]; however, Walker's axiom systems are not categorical, so his "completion" can not be justified by showing that it is trivial, as in the present treatment.

§3.1 Conditional Completion of a Particle

We will define some notions which apply to any linearly ordered set X, and to any particle, since particles are linearly order sets of instants. An element $a \in X$ is called the *first element* of X if, for all $x \in X$, $a \leq x$. Similarly, an element $b \in X$ is called the *last element* of X if, for all $x \in X$, $b \geq x$. If X can be decomposed into two non-empty subsets $X_{(1)}$, $X_{(2)}$ such that

(i) $X_{(1)} \cup X_{(2)} = X$ and

(ii) for all $x_1 \varepsilon X_{(1)}$ and for all $x_2 \varepsilon X_{(2)}$, $x_1 < x_2$, then the decomposition $[X_{(1)}, X_{(2)}]$ is called a *cut*. If $X_{(1)}$ has no last element and $X_{(2)}$ has no first element, we say that the cut $[X_{(1)}, X_{(2)}]$ is a *gap* (in X). An ordered set X is *conditionally complete* if every bounded subset has an infimum and a supremum.

Any linearly ordered set X can be embedded in a conditionally complete linearly ordered set $\overline{X}$, as is shown by MacNeille [1937, §11, Theorem 11.7] and also, in a form closer to the present treatment, by Birkhoff [1967, Chapter V, §9]. The adjoined elements correspond to those cuts which are gaps in X. The order relation on $\overline{X}$ is defined:

for any $y \varepsilon X$ and any gaps $[X_{(1)}, X_{(2)}]$, $[X_{(3)}, X_{(4)}]$,

$$y \varepsilon X_{(1)} \Leftrightarrow y < [X_{(1)}, X_{(2)}],$$
$$y \varepsilon X_{(2)} \Leftrightarrow y > [X_{(1)}, X_{(2)}],$$
$$X_{(1)} \subset X_{(3)} \Leftrightarrow [X_{(1)}, X_{(2)}] < [X_{(3)}, X_{(4)}].$$

The conditional completion of a particle $\underset{\sim}{Q}$ is denoted by $\overline{\underset{\sim}{Q}}$ and is called a *conditionally complete particle*. Instants of the given particle $\overline{\underset{\sim}{Q}}$ are called *ordinary instants*, while the adjoined elements are called *ideal instants*: all elements of the conditionally complete particle $\overline{\underset{\sim}{Q}}$ will from now on be called *instants*. Instants of $\overline{\underset{\sim}{Q}}$ are denoted by the symbol $\overline{Q}$ together with subscripts, for example, $\overline{Q}_1$, $\overline{Q}_a$, $\overline{Q}_\alpha$, $\overline{Q}_x \varepsilon \overline{\underset{\sim}{Q}}$. If $\overline{Q}_x \varepsilon \overline{\underset{\sim}{Q}}$, and if it is known that $\overline{Q}_x$ is an ordinary instant, we shall write Q_x instead of $\overline{Q}_x$; thus the statement "$Q_x \in \overline{\underset{\sim}{Q}}$" means that Q_x is an ordinary instant of $\overline{\underset{\sim}{Q}}$ (and of $\underset{\sim}{Q}$), and the statement "$\overline{Q}_x = Q_x \in \overline{\underset{\sim}{Q}}$" means that $\overline{Q}_x$ is an ordinary instant of $\overline{\underset{\sim}{Q}}$. *The set of all conditionally complete particles* is denoted by $\overline{\mathcal{P}}$. *The set of all instants* (ordinary and ideal) is denoted by $\overline{\mathcal{I}}$.

§3.2 Properties of Extended Signal Relations and Functions

Since ideal instants correspond to gaps, the Signal Axiom (Axiom I, §2.2) induces a binary relation $\overline{\sigma}$ on $\overline{\mathcal{I}} \times \overline{\mathcal{I}}$. The *extended signal relation* $\overline{\sigma}$ is a unique extension of σ on $\overline{\mathcal{I}} \times \overline{\mathcal{I}}$, and so we shall abbreviate the symbol $\overline{\sigma}$ to σ. Similarly, there are *extended signal functions*, an *extended coincidence relation*, *ideal events*, an *extended relation of "in optical line"*, and an *extended betweenness relation*, *all of which will be denoted by the previous symbols. The (extended) signal functions are bijections which send ordinary instants onto ordinary instants and ideal instants onto ideal instants.* The following theorem is essentially the same as a result of Walker [1948, P322].

THEOREM 8 (Extended Signal Functions)

Let $\overline{\underset{\sim}{Q}}, \overline{\underset{\sim}{R}}, \overline{\underset{\sim}{S}} \in \overline{\mathcal{P}}$. *Then* (i) f_{SQ} *is order-preserving and* (ii) $f_{SQ} \leq f_{SR} \circ f_{RQ}$.

This theorem is a consequence of Axiom IV (§2.4) and Theorem 1 (§2.5). It is used in the proof of Theorems 9 (§3.2), 11(§3.4), 12(§3.5), 14(§3.6) and Corollary 1 of Theorem 22 (§5.2).

PROOF (in outline)

It is only necessary to prove these results for ideal instants.
(i) Consider an ideal instant $\overline{Q}_y \in \overline{\underset{\sim}{Q}}$ and any other instant of $\overline{\underset{\sim}{Q}}$. Apply the (ordinary) signal function f_{SQ} to the definition of the order relation for a conditionally completed particle (see the previous section).
(ii) Given an ideal instant $\overline{Q}_y \in \overline{\underset{\sim}{Q}}$, apply the Triangle Inequality (Axiom IV, §2.4) to all ordinary instants of the cut (in $\underset{\sim}{Q}$) which corresponds to $\overline{Q}_y$. □

COROLLARY (Walker [1948], Theorem 5.5, P323)

Given ideal particles $\overline{\underset{\sim}{Q}}, \overline{\underset{\sim}{R}}$ *and instants* $\overline{Q}_x, \overline{Q}_z \in \overline{\underset{\sim}{Q}}$ *and* $\overline{R}_y \in \overline{\underset{\sim}{R}}$, *if* $\overline{Q}_x \; \sigma \; \overline{R}_y \; \sigma \; \overline{Q}_z$, *then* $\overline{Q}_x \simeq \overline{R}_y \simeq \overline{Q}_z = \overline{Q}_x$ *or* $Q_x < Q_z$.

PROOF. For ordinary instants, this is simply the definition of the temporal relation (§2.3). If $\overline{Q}_x$ is an ideal instant the definition of the temporal relation can be applied to all ordinary instants in the cut corresponding to $\overline{Q}_x$. Thus $\overline{Q}_x \leq \overline{Q}_z$. If $\overline{Q}_x = \overline{Q}_z$, then $\overline{Q}_x \; \sigma \; \overline{R}_y \; \sigma \; \overline{Q}_x \; \sigma \; \overline{R}_y$, that is $\overline{Q}_x \simeq \overline{R}_y$, which completes the proof. □

Statements similar to those contained in the following theorem have been made by Walker [1948, P322, P325]. The theorem is not stated in terms of the concept of a "limit", since we have not yet shown that a particle is a Hausdorff set (of instants). It would have been possible to insert Theorem 15 (of §3.7) at an earlier stage and thus simplify the statement of the next theorem; but it is not necessary to do so, and the present approach is independent of the Axiom of Incidence (Axiom IX, §2.11).

THEOREM 9 (Monotone Sequence Theorem)

Let $\underset{\sim}{\overline{Q}}, \underset{\sim}{\overline{R}} \in \overline{\mathcal{P}}$ *and let* $\underset{\sim}{Q}'$ *be a subset of* $\underset{\sim}{\overline{Q}}$.
If $\underset{\sim}{Q}'$ *is bounded above, it has a supremum and*

$$\sup\{f_{RQ}(\overline{Q}_t): \overline{Q}_t \in \underset{\sim}{Q}'\} = f_{RQ}(\sup\{\overline{Q}_t: \overline{Q}_t \in \underset{\sim}{Q}'\}) .$$

If $\underset{\sim}{Q}'$ *is bounded below, it has an infimum and*

$$\inf\{f_{RQ}(\overline{Q}_t): \overline{Q}_t \in \underset{\sim}{Q}'\} = f_{RQ}(\inf\{\overline{Q}_t: \overline{Q}_t \in \underset{\sim}{Q}'\}) .$$

In particular, if $\{\overline{Q}_n: n=1,2,\ldots;\ \overline{Q}_n \in \underset{\sim}{\overline{Q}}\}$ *is a bounded increasing sequence, then*

$$\sup\{f_{RQ}(\overline{Q}_n)\} = f_{RQ}(\sup\{\overline{Q}_n\}) .$$

If $\{\overline{Q}_n: n=1,2,..;\ \overline{Q}_n \in \underset{\sim}{\overline{Q}}\}$ *is a bounded decreasing sequence, then*

$$\inf f_{RQ}(\overline{Q}_n) = f_{RQ}(\inf\{\overline{Q}_n\}).$$

This theorem is a consequence of Theorem 8 (§3.2). It is used in the proof of Theorems 12 (§3.5), 17 (§4.2), Corollary 2 of Theorem 22 (§5.2), and Theorems 32 (§6.4), 36 (§7.1), 41 (§7.3), 42 (§7.3) and 49 (§7.5).

PROOF. By the previously quoted theorem of MacNeille [1937], $\underset{\sim}{Q}'$ has a supremum or an infimum, as the case may be. Since (extended) signal functions are one-to-one and order-preserving, by the previous theorem, f_{RQ} if an order-isomorphism between $\underset{\sim}{\overline{Q}}$ and $\underset{\sim}{\overline{R}}$, which leads to the stated equalities. □

§3.3 Generalised Triangle Inequalities

The next result is due to Walker [1948, Theorem 5.2, P323].

THEOREM 10. *Let $\underset{\sim}{\overline{Q}}, \underset{\sim}{\overline{R}}, \ldots, \underset{\sim}{\overline{T}}$ be a finite set of conditionally complete particles with instants $\overline{Q}_x, \overline{Q}_y \in \underset{\sim}{\overline{Q}}$; $\overline{R}_x, \overline{R}_y \in \underset{\sim}{\overline{R}}; \ldots$; and $\overline{T}_x, \overline{T}_y \in \underset{\sim}{\overline{T}}$. Then*

(i) *If $\overline{Q}_x \;\sigma\; \overline{R}_y \;\sigma\; \cdots \;\sigma\; \overline{T}_y$ and $\overline{Q}_x \;\sigma\; \overline{T}_x$, then $\overline{T}_x \leqslant \overline{T}_y$.*

(ii) *If $\overline{Q}_x \;\sigma\; \overline{R}_x \;\sigma\; \cdots \;\sigma\; \overline{T}_x$ and $\overline{Q}_y \;\sigma\; \overline{T}_x$, then $\overline{Q}_x \leqslant \overline{Q}_y$.*

This theorem is a consequence of Theorems 1 (§2.5) and 8 (§3.2). It is used in the proof of Theorems 13 (§3.6) and 17 (§4.2).

PROOF. Part (i) is a consequence of Theorem 8 (§3.2) by induction on the number of particles. Part (ii) follows from part (i) by Theorem 8(i). □

§3.4 Particles Do Not Have First or Last Instants

The next theorem applies to ideal as well as to ordinary instants. The first paragraph of the proof resembles a theorem of Walker [1948, Theorem 5.1, P323].

THEOREM 11. *Particles do not have first or last instants.*

This theorem is a consequence of Axioms II (§2.3) and VIII (§2.10) and Theorems 6 (§2.9) and 8 (§3.2). It is used in the proof of Theorem 25 (§5.4) and the Corollary to Theorem 40 (§7.3).

PROOF. By Theorem 8 (§3.2), if (conditionally complete) particles had first (or last) instants, they would all coincide at their first (or last) instants. Signal functions are order-preserving bijections, so if one particle has a first (or last) ordinary instant, then all particles have a first (or last) ordinary instant, and all particles would coincide at their first (or last) ordinary instants.

By the Axiom of Dimension (Axiom VIII, §2.10), there are two distinct particles $\underset{\sim}{Q}^{(1)}, \underset{\sim}{Q}^{(2)}$ which we shall denote by $\underset{\sim}{S}, \underset{\sim}{T}$ (for ease of writing) with instants $S_0 \in \underset{\sim}{S}$, $T_0 \in \underset{\sim}{T}$ such that $S_0 \simeq T_0$. Since $\underset{\sim}{S} \not\simeq \underset{\sim}{T}$, there is an instant $S_x \in \underset{\sim}{S}$ with $S_x \neq S_0$. By the First Axiom of Temporal Order (Axiom II, §2.3), there is a particle $\underset{\sim}{U}$ such that

(i) if $S_0 < S_x$, $f_{SU} \circ f_{US}(S_0) = S_x$, or

(ii) if $S_0 > S_x$, $f_{SU} \circ f_{US}(S_x) = S_0$.

Thus, in either case, respectively,

(i) $(f_{SU} \circ f_{US})^{-1}(S_0) < S_0$ and

(ii) $f_{SU} \circ f_{US}(S_0) > S_0$,

which shows that S_0 is neither the first instant nor the last instant of $\underset{\sim}{S}$. If $\underset{\sim}{S}$ had a first or last (ordinary) instant, then Theorem 6(§2.9) would imply that $\underset{\sim}{S} \simeq \underset{\sim}{T}$, which would be a contradiction.

Therefore no particle has a first (ordinary) instant or a last (ordinary) instant. By definition, gaps can not correspond to first or last instants, so no particle has a first or last ideal instant. □

§3.5 Events at Which Distinct Particles Coincide

THEOREM 12. *Given distinct particles* $\underset{\sim}{Q}, \underset{\sim}{T}$ *and an instant* $T_0 \in \underset{\sim}{T}$*, we define for all integers* n,

$$T_n \overset{def}{=} (f_{TQ} \circ f_{QT})^n (T_0).$$

If $\sup \{T_n\} \in \overline{\underset{\sim}{T}}$*, let* $\overline{T}_\infty \overset{def}{=} \sup\{T_n\}$*, and if* $\inf\{T_n\} \in \overline{\underset{\sim}{T}}$*, let* $\overline{T}_{-\infty} \overset{def}{=} \inf\{T_n\}$*. Then*

(i) *if* $\overline{T}_\infty$ *exists,* $\overline{\underset{\sim}{Q}}$ *coincides with* $\overline{\underset{\sim}{T}}$ *at* $[\overline{T}_\infty]$,

(ii) *if* $\overline{T}_{-\infty}$ *exists,* $\overline{\underset{\sim}{Q}}$ *coincides with* $\overline{\underset{\sim}{T}}$ *at* $[\overline{T}_{-\infty}]$*, and*

(iii) *for all* $T_x \in \bigcup_{n=-\infty}^{+\infty} \{\overline{T}_y : T_n \leq \overline{T}_y \leq T_{n+1}, \overline{T}_y \in \overline{\underset{\sim}{T}}\}$*,* $\overline{\underset{\sim}{Q}}$ *does not coincide with* $\overline{\underset{\sim}{T}}$ *at* $[\overline{T}_x]$ (see Fig. 8).

This theorem is a consequence of Theorems 8 (§3.2) and 9 (§3.2). It is used in the proof of Theorems 15 (§3.7), 25 (§5.4), 40 (§7.3), 46 (§7.5) and 50 (§8.1).

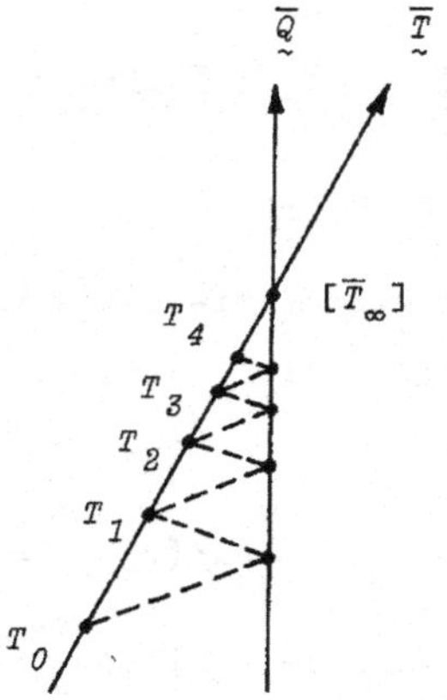

Fig. 8. In this illustration, $sup\{T_n\} \in \underset{\sim}{\overline{T}}$, and the particles $\underset{\sim}{\overline{Q}}$ and $\underset{\sim}{\overline{T}}$ coincide at the event $[\overline{T}_\infty]$.

PROOF. We define a function

$$q \overset{def}{=} f_{TQ} \circ f_{QT} ;$$

and then

$$T_n = q^n(T_0).$$

Consider an instant $\overline{T}_x \in \underset{\sim}{\overline{T}}$ with

$$T_0 < \overline{T}_x \leq T_1 ;$$

then, since q is a strictly monotone increasing function by

Theorem 8 (§3.2),

$$T_1 = q(T_0) < q(\overline{T}_x) \quad \text{and so}$$

$$\overline{T}_x < q(\overline{T}_x).$$

That is, for any instants $\overline{Q}_y \in \overline{\underset{\sim}{Q}}$ and $\overline{T}_x \in \overline{\underset{\sim}{T}}$ with $T_0 < \overline{T}_x \leq T_1$,

$$\overline{T}_x \; \sigma \; \overline{Q}_y \Rightarrow \overline{Q}_y \not\sigma \; \overline{T}_x.$$

Similar considerations apply for each semi-closed interval $T_n < \overline{T}_x \leq T_{n+1}$, which proves (iii). If $\overline{T}_\infty$ exists, Theorem 9 (§3.2) implies that

$$q(\overline{T}_\infty) = q(sup\{T_n\}) = sup\{q(T_n)\} = sup\{T_{n+1}\} = \overline{T}_\infty \; ,$$

which proves (i). The proof of (ii) is similar. □

This theorem and its proof are based on a theorem of Walker [1948, Theorem 8.1, P325]. Walker has paraphrased the result in the following words: "any instant of $\underset{\sim}{T}$, at which $\underset{\sim}{T}$ does not coincide with $\underset{\sim}{Q}$, lies within an interval of $\underset{\sim}{T}$ at every instant of which $\underset{\sim}{T}$ does not coincide with $\underset{\sim}{Q}$".

§3.6 Generalised Temporal Order. Relations on the Set of Events. Observers.

An intuitively obvious property of the coincidence relation is proved in the following theorem. This allows us to generalise the signal relation and related concepts to apply to the set of events. Walker [1948, Theorem 5.4, P323] proved a special case of the following theorem, and generalised the temporal order relation to apply to instants of different particles; however he did not relate instants and events in either of his papers (Walker [1948, 1959]).

THEOREM 13 (Substitution Property of the Coincidence Relation).

Let $\underset{\sim}{\overline{Q}}, \underset{\sim}{\overline{R}}, \underset{\sim}{\overline{S}}, \underset{\sim}{\overline{T}}$ *be conditionally complete particles with instants* $\overline{Q}_1 \in \underset{\sim}{\overline{Q}}, \overline{R}_2 \in \underset{\sim}{\overline{R}}, \overline{S}_3 \in \underset{\sim}{\overline{S}}, \overline{T}_4 \in \underset{\sim}{\overline{T}}$.
If $\overline{Q}_1 \;\sigma\; \overline{R}_2 \;\sigma\; \overline{T}_3$ *and* $\overline{R}_2 \simeq \overline{S}_2$, *then* $\overline{Q}_1 \;\sigma\; \overline{S}_2 \;\sigma\; \overline{T}_3$.

This theorem is a consequence of Axiom I (§2.2) and Theorem 10 (§3.3). It is used in the proofs of Theorems 14 (§3.6) and 16 (§4.1).

PROOF By the Signal Axiom (Axiom I, §2.2) there are instants $\overline{Q}_x \in \underset{\sim}{\overline{Q}}$ and $\overline{T}_y \in \underset{\sim}{\overline{T}}$ such that

$$\overline{Q}_x \;\sigma\; \overline{S}_2 \;\sigma\; \overline{T}_y .$$

So we have

$$\overline{Q}_1 \;\sigma\; \overline{R}_2 \;\sigma\; \overline{S}_2 \text{ and } \overline{Q}_x \;\sigma\; \overline{S}_2 \text{ and}$$
$$\overline{Q}_x \;\sigma\; \overline{S}_2 \;\sigma\; \overline{R}_2 \text{ and } \overline{Q}_1 \;\sigma\; \overline{R}_2 ,$$

which, by the Generalised Triangle Inequalities (Theorem 10 §3.3), imply that $\overline{Q}_1 \leq \overline{Q}_x$ and $\overline{Q}_x \leq \overline{Q}_1$, respectively. So $\overline{Q}_x = \overline{Q}_1$; that is, $\overline{Q}_1 \;\sigma\; \overline{S}_2$.

Similarly $\overline{R}_2 \ \sigma \ \overline{S}_2 \ \sigma \ \overline{T}_y$ and $\overline{R}_2 \ \sigma \ \overline{T}_3$,

and $\overline{S}_2 \ \sigma \ \overline{R}_2 \ \sigma \ \overline{T}_3$ and $\overline{S}_2 \ \sigma \ \overline{T}_y$

imply that $\overline{T}_y = \overline{T}_3$; that is, $\overline{S}_2 \ \sigma \ \overline{T}_3$. □

COROLLARY *If* $\underset{\sim}{R} \simeq \underset{\sim}{S}$, *then* $f_{QR} \circ f_{RQ} = f_{QS} \circ f_{SQ}$. □

Since (extended) signal functions are order-preserving (Theorem 8, §3.2), we can *now define a generalisation of the (extended) temporal order relation* for pairs of instants which need not belong to the same particle. Given two conditionally complete particles $\underset{\sim}{\overline{Q}},\underset{\sim}{\overline{R}}$ with instants $\overline{Q}_x \in \underset{\sim}{\overline{Q}}$ and $\overline{R}_x \in \underset{\sim}{\overline{R}}$, we say that $\overline{Q}_x$ is *before* $\overline{R}_y$ and $\overline{R}_y$ is *after* $\overline{Q}_x$ if:

(i) $f_{RQ}(\overline{Q}_x) < \overline{R}_y$, or

(ii) $f_{RQ}(\overline{Q}_x) = \overline{R}_y$ and $\overline{Q}_x \neq \overline{R}_y$.

If $\overline{Q}_x$ is before $\overline{R}_y$ we write $\overline{Q}_x < \overline{R}_y$ and if $\overline{R}_y$ is after $\overline{Q}_x$ we write $\overline{R}_y > \overline{Q}_x$, as in §2.3. Note that there could be pairs of instants between which none of the relations $<$, $\simeq$, or $>$ hold. The above definition is a paraphrased version of a similar definition of Walker [1948, P323].

A particular case of the (generalised) temporal order relation occurrs for instants which are in optical line. Given particles $\underset{\sim}{\overline{Q}},\underset{\sim}{\overline{R}},\underset{\sim}{\overline{S}}$ and non-coincident instants $\overline{Q}_x \in \underset{\sim}{\overline{Q}}$, $\overline{R}_y \in \underset{\sim}{\overline{R}}$, $\overline{S}_z \in \underset{\sim}{\overline{S}}$ such that $|\overline{Q}_x,\overline{R}_y,\overline{S}_z>$,
the instant $\overline{Q}_x$ is before the instants $\overline{R}_y$ and $\overline{S}_z$,
the instant $\overline{S}_z$ is after the instants $\overline{Q}_x$ and $\overline{R}_y$, and the instant $\overline{R}_y$ is after $\overline{Q}_x$ and before $\overline{S}_z$.

THEOREM 14. *Let* $\underset{\sim}{\overline{Q}}, \underset{\sim}{\overline{R}}, \underset{\sim}{\overline{S}}$ *be conditionally complete particles with instants* $\overline{Q}_x \in \underset{\sim}{\overline{Q}}$, $\overline{R}_y \in \underset{\sim}{\overline{R}}$, *and* $\overline{S}_z \in \underset{\sim}{\overline{S}}$. *Then*

(i) $\overline{Q}_x < \overline{R}_y \Rightarrow \overline{Q}_x \not\simeq \overline{R}_y$ *and* $\overline{Q}_x \not> \overline{R}_y$,

(ii) $\overline{Q}_x \simeq \overline{R}_y \Rightarrow \overline{Q}_x \not< \overline{R}_y$ *and* $\overline{Q}_x \not> \overline{R}_y$,

(iii) $\overline{Q}_x < \overline{R}_y$ *and* $\overline{R}_y < \overline{S}_z \Rightarrow \overline{Q}_x < \overline{S}_z$,

(iv) $\overline{Q}_x < \overline{R}_y$ *and* $\overline{R}_y \simeq \overline{S}_z \Rightarrow \overline{Q}_x < \overline{S}_z$,

(v) $\overline{Q}_x \simeq \overline{R}_y$ *and* $\overline{R}_y < \overline{S}_z \Rightarrow \overline{Q}_x < \overline{S}_z$.

This theorem is essentially the same as a theorem of Walker [1948, Theorem 5.6, P323]. It is a consequence of Theorems 8 (§3.2) and 13 (§3.5), and is used in the proof of Theorem 32 (§6.4).

PROOF. By Theorem 8 (§3.2) and Theorem 13. □

The substitution property of the coincidence relation (Theorem 13, §3.6) permits the σ-relation to be generalised so as to apply to events and hence *the signal functions, the temporal order relation, and the concept of optical lines can all be extended to apply to the set of events.*

Given events $[\overline{Q}_x], [\overline{R}_y], [\overline{S}_z]$ we define:

(i) $[\overline{Q}_x] \; \sigma \; [\overline{R}_y]$ if and only if there are instants $\overline{T}_x \in [\overline{Q}_x]$ and $\overline{U}_y \in [\overline{R}_y]$ such that $\overline{T}_x \; \sigma \; \overline{U}_y$;

(ii) $f_{RQ} [\overline{Q}_x] = [\overline{R}_y]$ if and only if $[\overline{Q}_x] \; \sigma \; [\overline{R}_y]$;

(iii) $|[\overline{Q}_x],[\overline{R}_y],[\overline{S}_z]>$ if and only if there are instants $\overline{T}_x \in [\overline{Q}_x]$, $\overline{U}_y \in [\overline{R}_y]$, and $\overline{V}_z \in [\overline{S}_z]$ such that $|\overline{T}_x, \overline{U}_y, \overline{V}_z>$; and

(iv) $[\overline{Q}_x] < [\overline{R}_y]$ if and only if there are instants $\overline{T}_x \in [\overline{Q}_x]$ and $\overline{U}_y \in [\overline{R}_y]$ such that $\overline{T}_x < \overline{U}_y$.

It is a consequence of the previous theorem that *the temporal order relation is transitive on the set of events.*

An important consequence of Theorem 13 is that any composition of signal functions is unaltered by changing any given particle to a particle permanently coincident with the given particle, provided that the domain and range of the composition is unaltered; that is,

$$\underset{\sim}{R} \simeq \underset{\sim}{S} \Longrightarrow \cdots f_{QR} \circ f_{RT} \cdots = \cdots f_{QS} \circ f_{ST} \cdots .$$

To each particle $\underset{\sim}{Q}$ we define a corresponding *observer:*

$$\hat{\underset{\sim}{Q}} \overset{def}{=} \{\underset{\sim}{R} : \underset{\sim}{R} \simeq \underset{\sim}{Q}, \underset{\sim}{R} \in \mathcal{P}\} .$$

We see from the above remarks that particles belonging to the same observer "appear to be the same" as "seen by any another particle"; that is, if $\underset{\sim}{R}, \underset{\sim}{S} \in \hat{\underset{\sim}{Q}}$ then for any particle $\underset{\sim}{T}$,

$$f_{TR} \circ f_{RT} = f_{TS} \circ f_{ST} .$$

Observers have been defined as equivalence classes of coincident particles, which is analogous to the definition of events as equivalence classes of coincident instants. We do not define "conditionally complete observers" since we have no use for such a concept. (In §5.4 we will show that all ordinary particles are conditionally complete, which means that the previous completion is trivial).

Several definitions which apply to particles can now be extended to observers: for example:

(i) $f_{RQ} : \hat{\underset{\sim}{Q}} \to \hat{\underset{\sim}{R}}$

$[Q_x] \mapsto [R_y]$ if and only if $[Q_x]\,\sigma\,[R_y]$.

(ii) $[\hat{\underset{\sim}{Q}}, \hat{\underset{\sim}{R}}, \hat{\underset{\sim}{S}}] \leftrightarrow$ for all $\underset{\sim}{T} \in \hat{\underset{\sim}{Q}}$, for all $\underset{\sim}{U} \in \hat{\underset{\sim}{R}}$, and for all $\underset{\sim}{V} \in \underset{\sim}{S}$, $[\underset{\sim}{T}, \underset{\sim}{U}, \underset{\sim}{V}]$.

(iii) $\langle\hat{\underset{\sim}{Q}}, \hat{\underset{\sim}{R}}, \hat{\underset{\sim}{S}}\rangle \leftrightarrow$ for all $\underset{\sim}{T} \in \hat{\underset{\sim}{Q}}$, for all $\underset{\sim}{U} \in \hat{\underset{\sim}{R}}$, and for all $\underset{\sim}{V} \in \hat{\underset{\sim}{S}}$, $\langle\underset{\sim}{T}, \underset{\sim}{U}, \underset{\sim}{V}\rangle$.

These definitions are consistent with the previous definitions which extended the signal relation and the relation "in optical line" to apply to the set of events.

§3.7 Each Particle is Dense in Itself

Let X be a linearly ordered set. If $Y \subseteq X$ and if, for all $x_1, x_2 \in X$ with $x_1 < x_2$, there exists some $y \in Y$ such that $x_1 < y < x_2$, we say that Y is a *dense subset* of X. If X is a dense subset of X, we say that X is *dense in itself*, or simply that X is a *dense set*.

THEOREM 15 (Each Particle is Dense in Itself)

Given a particle $\underset{\sim}{Q}$ *and ordinary instants* $Q_a, Q_c \in \underset{\sim}{Q}$ *with* $Q_a < Q_c$, *there is an instant* $Q_b \in \underset{\sim}{Q}$ *such that* $Q_a < Q_b < Q_c$.

This theorem is a consequence of Axioms VIII (§2.10), IX (§2.11) and X (§2.12) and Theorem 12 (§3.5). It is used in the proof of Theorems 23 (§5.3) and 40 (§7.3).

PROOF. By the Axiom of Dimension (Axiom VIII, §2.10) and the Axiom of Connectedness (Axiom X, §2.12), there is some particle distinct from $\underset{\sim}{Q}$, which coincides with $\underset{\sim}{Q}$ at some event. If this event is not $[Q_c]$, the Axiom of Incidence (Axiom IX, §2.11) implies that there is some particle $\underset{\sim}{S}$, distinct from $\underset{\sim}{Q}$, which coincides with $\underset{\sim}{Q}$ at $[Q_c]$. By Theorem 12 (§3.5),

$$Q_a < Q_b \overset{def}{=} f_{QS} \circ f_{SQ}(Q_a) < Q_c \ . \quad \square$$

COROLLARY. *Each conditionally complete particle is dense in itself. Moreover, each particle is a dense subset of its conditional completion.*

PROOF Let $\underset{\sim}{\overline{Q}}$ be a conditionally complete particle with instants $\overline{Q}_a, \overline{Q}_c \in \underset{\sim}{\overline{Q}}$ such that $\overline{Q}_a < \overline{Q}_c$.

Case 1. If $\overline{Q}_a$ or $\overline{Q}_c$ (or both) are ideal, then by §3.1, there is some ordinary instant $Q_b \in \underset{\sim}{Q}$ with $\overline{Q}_a < Q_b < \overline{Q}_c$.

Case 2. If both $\overline{Q}_a$ and $\overline{Q}_c$ are ordinary instants, the above theorem applies. $\square$

CHAPTER 4

IMPLICATIONS OF COLLINEARITY

Most of the results contained in this chapter have been given by Walker [1948] but since the present axiom system differs from Walker's, proofs have been given in detail for the sake of logical completeness.

§4.1 Collinearity. The Two Sides of an Event.

A set of particles Σ is *collinear* if, for all particles $\underset{\sim}{Q} \in \Sigma$ and for each instant $Q_x \in \underset{\sim}{Q}$, either:

(i) there are two distinct optical lines, each containing Q_x and one instant from each particle of $\Sigma \setminus \{\underset{\sim}{Q}\}$, or

(ii) all particles of Σ coincide with $\underset{\sim}{Q}$ at $[Q_x]$.

We shall indicate that a set of particles is collinear by enclosing the particles in square brackets; thus $[\underset{\sim}{Q},\underset{\sim}{R},\underset{\sim}{S},\underset{\sim}{T}]$ means that $\{\underset{\sim}{Q},\underset{\sim}{R},\underset{\sim}{S},\underset{\sim}{T}\}$ is collinear. *The symbol Σ will be used to denote an arbitrary collinear set of particles.*

Before establishing the main result we prove the following:

PROPOSITION (Walker [1948], Theorem 7.2, P324)

Let $\underset{\sim}{Q}, \underset{\sim}{S}, \underset{\sim}{T} \in \Sigma$ *and let* $S_y \in \underset{\sim}{S}$.

If $|f^{-1}_{SQ}(S_y), S_y, f_{TS}(S_y)>$, *then* $|f^{-1}_{ST}(S_y), S_y, f_{QS}(S_y)>$.

That is, the instant S_y *is between* $\underset{\sim}{Q}$ *and* $\underset{\sim}{T}$ *and by Theorem 13 (§3.6), the event* $[S_y]$ *is between* $\underset{\sim}{Q}$ *and* $\underset{\sim}{T}$.

PROOF. Let $Q_x \overset{def}{=} f^{-1}_{SQ}(S_y)$, $T_x \overset{def}{=} f^{-1}_{ST}(S_y)$, $Q_z \overset{def}{=} f_{QS}(S_y)$ and $T_z \overset{def}{=} f_{TS}(S_y)$. We must show that $|Q_x, S_y, T_z>$ implies $|T_x, S_y, Q_z>$. Consider the optical line which contains the instants T_x and S_y . If $T_x \simeq S_y$, there is nothing further to prove. If $T_x \not\simeq S_y$, then the instant of $\underset{\sim}{Q}$ which is in optical line with T_x and S_y is either:

(i) Q_x which implies $|Q_x, T_x, S_y>$ or $|T_x, Q_x, S_y>$, or

(ii) Q_z which implies $|T_x, S_y, Q_z>$.

Now $|Q_x, T_x, S_y>$, the data, and Theorem 3 (§2.7) imply $|Q_x, T_x, S_y, T_z>$ and by the Signal Axiom (Axiom I, §2.2), $T_x = T_z$ so $T_x \simeq S_y$, which is a contradiction. Also $|T_x, Q_x, S_y>$ and the Axiom of Uniqueness of Extension of Optical Lines (Axiom V, §2.7) imply that $|T_x, Q_x, S_y, T_z>$ and by Theorem 7 (§2.12), $T_x = T_z$, so $T_x \simeq S_y$, which is another contradiction. The only remaining possibility is (ii) above, which was the result to be proved. □

The proposition can now be extended:

THEOREM 16 (Walker [1948], Theorem 7.3, P324)

Given a particle $\underset{\sim}{S} \in \Sigma$ *and an instant* $S_2 \in \underset{\sim}{S}$*, each particle of* Σ *can be placed in one of three disjoint subsets* $\mathcal{L}[S_2]$, $\mathcal{C}[S_2]$, $\mathcal{R}[S_2] \subset \Sigma$. *Particles in* $\mathcal{C}[S_2]$ *coincide with* $\underset{\sim}{S}$ *at* $[S_2]$*; the event* $[S_2]$ *is between any particle of* $\mathcal{L}[S_2]$ *and any particle of* $\mathcal{R}[S_2]$*, but not between any two particles of* $\mathcal{L}[S_2]$ *or of* $\mathcal{R}[S_2]$*. The sets of particles* $\mathcal{L}[S_2]$ *and* $\mathcal{R}[S_2]$ *are called the left side (of* $[S_2]$ *in* Σ*) and the right side (of* $[S_2]$ *in* Σ*), respectively.*

This theorem is a consequence of Axioms I (§2.2) and V (§2.7) and Theorems 3 (§2.7), 7 (§2.12) and 13 (§3.6). It is used in the proof of Theorem 17 (§4.2).

PROOF. If all particles in Σ coincide at $[S_2]$, there is nothing further to prove. Otherwise there is a particle $\underset{\sim}{T} \in \Sigma$ such that T does not coincide with S at $[S_2]$. By the Signal Axiom (Axiom I, §2.2) there are instants $T_0, T_4 \in \underset{\sim}{T}$ such that $T_0 \;\sigma\; S_2 \;\sigma\; T_4$. Again by the Signal Axiom, for any particles $\underset{\sim}{Q}, \underset{\sim}{U} \in \Sigma$ there are instants:

$$Q_1, Q_3 \in \underset{\sim}{Q} \text{ such that } Q_1 \;\sigma\; S_2 \;\sigma\; Q_3 \text{ and}$$

$$U_1, U_3 \in \underset{\sim}{U} \text{ such that } U_1 \;\sigma\; S_2 \;\sigma\; U_3 \;.$$

We specify that $\underset{\sim}{T} \in \mathcal{R}[S_2]$, so by the previous proposition:
for any $\underset{\sim}{Q} \in \Sigma$ such that $|Q_1, S_2, T_4\rangle$ and $Q_1 \neq Q_3$, $\underset{\sim}{Q} \in \mathcal{L}[S_2]$;
for any $\underset{\sim}{U} \in \Sigma$ such that $|S_2, U_3, T_4\rangle$ or $|S_2, T_4, U_3\rangle$ and $U_1 \neq U_3$
$\underset{\sim}{U} \in \mathcal{R}[S_2]$; and
all other particles in Σ are in $\mathcal{E}[S_2]$.
Having specified $\mathcal{L}[S_2]$, $\mathcal{E}[S_2]$, and $\mathcal{R}[S_2]$, the remaining part of the theorem is a consequence of the previous proposition and the definition of Σ. □

§4.2 The Intermediate Instant Theorem

The next theorem is due to Walker [1948, Theorem 7.4, P324]. In the present treatment it is called the "Intermediate Instant Theorem" because of its resemblance to the "Intermediate Value Theorem" of real variable theory. Before proving this result we establish the following:

PROPOSITION (Walker [1948, Lemma, P325]).

Let $\underset{\sim}{T}, \underset{\sim}{U}, \underset{\sim}{V} \in \Sigma$ *and let* $T_a \in \underset{\sim}{T}$. *Let the function* $g\colon\ \overline{\underset{\sim}{T}} \to \overline{\underset{\sim}{T}}$ *be defined such that, for each instant* $\overline{T}_x \in \overline{\underset{\sim}{T}}$,

$$g(\overline{T}_x) \overset{def}{=} \min\{f_{TU} \circ f_{UT}(\overline{T}_x)\ ,\ f_{TV} \circ f_{VT}(\overline{T}_x)\}$$

If $\underset{\sim}{U}$ *and* $\underset{\sim}{V}$ *are on the same side of* $[T_a]$ *then, for all* $T_x \in \underset{\sim}{T}$ *with* $T_a \leqslant T_x \leqslant g(T_a)$, $\underset{\sim}{U}$ *and* $\underset{\sim}{V}$ *are on the same side of* $[T_x]$.

PROOF (See Fig. 9)

By the Signal Axiom (Axiom I, §2.2) there is an instant $U_b \in \underset{\sim}{U}$ such that $T_a \ \sigma \ U_b$. We assume, without loss of generality, that $[U_b]$ is between $\underset{\sim}{T}$ and $\underset{\sim}{V}$, or that $\underset{\sim}{U}$ coincides with $\underset{\sim}{V}$ at $[V_b]$. Then by the Signal Axiom (Axiom I, §2.2) and the previous theorem, there are instants $V_a, V_b \in \underset{\sim}{V}$ and $T_b, T_c \in \underset{\sim}{T}$ such that

$$|T_a, U_b, V_b\rangle \text{ and } |V_a, U_b, T_b\rangle \text{ and } V_b \ \sigma \ T_c \ .$$

By Theorem 10 (§3.3),

$U_b \ \sigma \ V_b \ \sigma \ T_c$ and $U_b \ \sigma \ T_b \Longrightarrow T_b \leqslant T_c$, so $g(T_a) = T_b$.

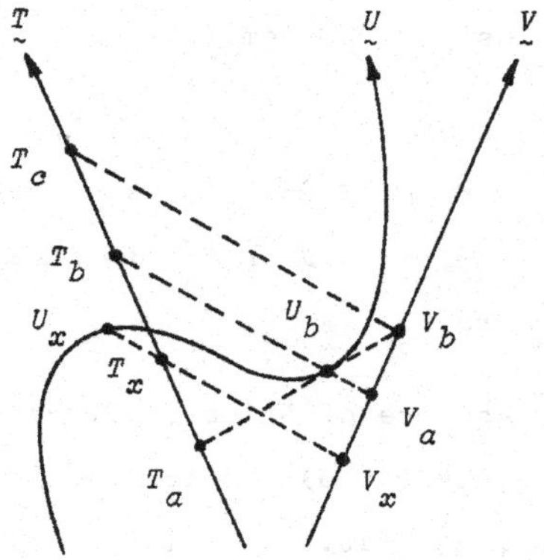

Fig. 9

We now suppose the contrary to the proposition; that is, we suppose that for some instant $T_x \in \underset{\sim}{T}$ with

$$T_a < T_x \leqslant T_b \ ,$$

$\underset{\sim}{U}$ and $\underset{\sim}{V}$ are on opposite sides of $[T_x]$; that is, we suppose that there are instants $U_x \in \underset{\sim}{U}$ and $V_x \in \underset{\sim}{V}$ such that

$$|V_x, T_x, U_x>$$

and we shall deduce a contradiction.

By Theorem 10 (§3.3),
$T_a \; \sigma \; U_b$ and $T_x \; \sigma \; U_x$ and $T_a < T_x \Rightarrow U_b < U_x$, so
$V_a \; \sigma \; U_b$ and $V_x \; \sigma \; U_x$ and $U_b < U_x \Rightarrow V_a < V_x$, but
$V_x \; \sigma \; T_x$ and $V_a \; \sigma \; T_b$ and $T_x \leqslant T_b \Rightarrow V_x \leqslant V_a$,
which is a contradiction. □

We can now prove:

THEOREM 17 (Intermediate Instant Theorem)

Let $\underset{\sim}{T}, \underset{\sim}{U}, \underset{\sim}{V} \in \Sigma$ *and let* $T_a, T_d \in \underset{\sim}{T}$.
If $\underset{\sim}{U}$ *and* $\underset{\sim}{V}$ *are on the same side of* $\underset{\sim}{T}$ *at* $[T_a]$*, and* $\underset{\sim}{U}$ *and* $\underset{\sim}{V}$ *are on opposite sides of* $\underset{\sim}{T}$ *at* $[T_d]$*, then* $\overline{\underset{\sim}{U}}$ *or* $\overline{\underset{\sim}{V}}$ *coincides with* $\overline{\underset{\sim}{T}}$ *at some instant between* T_a *and* T_d.

This theorem is a consequence of Axiom I (§2.2) and Theorems 1 (§2.5), 9 (§3.2), 10 (§3.3) and 16 (§4.1). It is used in the proof of Theorems 26 (§5.5), 28 (§6.1), 29 (§6.2), Corollary 1 to Theorem 30 (§6.3), and Theorems 33 (§6.4), 36 (§7.1) and 38 (§7.2).

PROOF. (Walker [1948, P325])

In the following argument we assume that $T_a < T_d$; the other case (with $T_a > T_d$) can be treated similarly. The function g is defined as in the preceding proposition. For any positive integer n, we let $T_n \overset{def}{=} g^n(T_a)$. The proposition could have been stated more generally, so as to apply to each closed interval $[T_n, T_{n+1}]$. Thus $\underset{\sim}{U}, \underset{\sim}{V}$ are on the same side of $\underset{\sim}{T}$ in the interval $\bigcup_{n=0}^{\infty} [T_n, T_{n+1}]$, and so for all n, $T_n < T_d$. By the Monotonic Sequence Theorem (Theorem 9, §3.2) there is an instant $\overline{T}_\infty \overset{def}{=} \sup_n\{T_n\} \in \overline{T}$ and $g(\overline{T}_\infty) = \overline{T}_\infty$, so $\underset{\sim}{\overline{T}}$ coincides with $\underset{\sim}{\overline{U}}$ or $\underset{\sim}{\overline{V}}$ at $[\overline{T}_\infty]$. Thus, if $\underset{\sim}{U}, \underset{\sim}{V}$ are on opposite sides of $\underset{\sim}{T}$ at $T_d > T_a$, then $T_d > \overline{T}_\infty > T_a$.
If $T_d < T_a$, the proof is similar. □

Thus, the two sides of $\underset{\sim}{T}$ in Σ are distinct until a particle from either side coincides with $\underset{\sim}{\overline{T}}$. In this interval, which could be a null interval, one side can be called the *left side* (of $\underset{\sim}{T}$) and the other can be called the *right side* (c.f. the previous section). In the following chapter we will be considering sub-SPRAYs which are collinear after the event of coincidence, and so the two sides of any particle in the sub-SPRAY can be well-defined after the event of coincidence. It is shown in Chapter 6 (§6.4) that collinear sets of particles exist and that left and right sides can be defined for (the set of all instants of) each particle.

An optical line containing instants S_x and T_y, where $T_y \in \underset{\sim}{T} \in \mathcal{R}[S_x]$, such that $S_x \sigma T_y$, is called a *right optical line* (through S_x). There is a similar definition for a *left optical line* (through S_x). In order that a right (left) optical line through S_x should be a right (left) optical line through T_y, we can define the sides of $[T_y]$ in Σ such that: for any particle $\underset{\sim}{U} \in \Sigma$ having an instant $U_z \in \underset{\sim}{U}$ such that $|S_x, T_y, U_z\rangle$ we define $\underset{\sim}{U} \in \mathcal{R}[T_y]$. Similarly, we can define $\mathcal{L}[T_y]$.

If $\underset{\sim}{U} \in \mathcal{R}[T_y]$ we say that $\underset{\sim}{U}$ is *to the right of* $[T_y]$; and similarly, if $\underset{\sim}{U} \in \mathcal{L}[T_y]$ we say that $\underset{\sim}{U}$ is *to the left of* $[T_y]$. If the instants S_x, T_y, U_z are on a right optical line such that

$$|S_x, T_y, U_z\rangle$$

we say that T_y is *to the right of* S_x and U_z is *to the right of* S_x and T_y; also T_y is *to the left of* U_z and S_x is *to the left of* T_y and U_z. We make similar definitions for the instants on a left optical line. *Along any given optical line, the relations "to the right of" and "to the left of" are transitive.*

§4.3 Modified Signal Functions and Modified Record Functions

We now derive some results which are used to simplify proofs in later sections. Walker [1948, §9, P326] has previously defined the "modified record function"; we define two kinds of "modified signal functions" in an analogous way.

Given two particles $\underset{\sim}{Q}, \underset{\sim}{R}$ in a collinear set, the *modified record function* $\left(f_{QR} \circ f_{RQ}\right)^*$ is defined:

$$\left(f_{QR} \circ f_{RQ}\right)^*(Q_x) \overset{def}{=} \begin{cases} f_{QR} \circ f_{RQ}(Q_x) & \text{, if } \underset{\sim}{R} \text{ is to the right of } Q_x, \\ Q_x & \text{, if } \underset{\sim}{R} \text{ coincides with } \underset{\sim}{Q} \text{ at } [Q_x], \\ \left(f_{QR} \circ f_{RQ}\right)^{-1}(Q_x) & \text{, if } \underset{\sim}{R} \text{ is to the left of } Q_x. \end{cases}$$

The modified record function indicates relative position, for

$$\left(f_{QR} \circ f_{RQ}\right)^*(Q_x) \gtrless Q_x$$

depending on whether $\underset{\sim}{R}$ is to the right, or left, of $[Q_x]$.

We define the *modified signal function* f^+_{RQ}, which is related to right optical lines, as follows:

$$f^+_{RQ}(Q_x) \overset{def}{=} \begin{cases} f_{RQ}(Q_x) & \text{, if } f_{RQ}(Q_x) \text{ is to the right of } Q_x, \\ Q_x & \text{, if } \underset{\sim}{R} \text{ coincides with } \underset{\sim}{Q} \text{ at } [Q_x], \text{ or} \\ f^{-1}_{QR}(Q_x) & \text{, if } f^{-1}_{QR}(Q_x) \text{ is to the left of } Q_x. \end{cases}$$

Similarly, we define the *modified signal function* f^-_{RQ}, which is related to left optical lines, as follows:

$$f^-_{RQ}(Q_x) \overset{def}{=} \begin{cases} f_{RQ}(Q_x) & \text{, if } f_{RQ}(Q_x) \text{ is to the left of } Q_x, \\ Q_x & \text{, if } \underset{\sim}{R} \text{ coincides with } \underset{\sim}{Q} \text{ at } [Q_x], \text{ or} \\ f^{-1}_{QR}(Q_x) & \text{, if } f^{-1}_{QR}(Q_x) \text{ is to the right of } Q_x. \end{cases}$$

THEOREM 18. *Let* $\underset{\sim}{Q}, \underset{\sim}{R}, \underset{\sim}{S} \in \Sigma$. *Then*

(i) $$\left(f^{+}_{RQ}\right)^{-1} = f^{+}_{QR}\ , \quad \left(f^{-}_{RQ}\right)^{-1} = f^{-}_{QR}\ ,$$

(ii) $$f^{+}_{SR} \circ f^{+}_{RQ} = f^{+}_{SQ}\ , \quad f^{-}_{SR} \circ f^{-}_{RQ} = f^{-}_{SQ} \quad \text{and}$$

(iii) $$\left(f_{QR} \circ f_{RQ}\right)^{*} = f^{-}_{QR} \circ f^{+}_{RQ}\ .$$

This theorem is a consequence of the previous definitions and is used in the proof of Theorems 33 (§6.4), 41 (§7.3), 43 (§7.4), 45 (§7.4), 48 (§7.5), 49 (§7.5), 50 (§8.1), 51 (§8.1), 52 (§8.2), 53 (§8.2), 54 (§8.3) and 55 (§8.3).

PROOF. Results (i) and (ii) are consequences of the previous definitions.

To establish (iii), we consider separately the possibilities of $\underset{\sim}{R}$ being to the right of $\underset{\sim}{Q}$, coincident with $\underset{\sim}{Q}$ (which is not shown since it is trivial), and to the left of $\underset{\sim}{Q}$. We apply the previous definitions. Thus

$$\left(f_{QR} \circ f_{RQ}\right) = \begin{cases} f_{QR} \circ f_{RQ} \\ \left(f_{QR} \circ f_{RQ}\right)^{-1} \end{cases}$$

$$= \begin{cases} f_{QR} \circ f_{RQ} \\ \left(f_{RQ}\right)^{-1} \circ \left(f_{QR}\right)^{-1} \end{cases}$$

$$= \begin{cases} f_{QR} \\ \left(f_{RQ}\right)^{-1} \end{cases} \circ \begin{cases} f_{RQ} \\ \left(f_{QR}\right)^{-1} \end{cases} = f^{-}_{QR} \circ f^{+}_{RQ}\ . \qquad \square$$

THEOREM 19 (Walker [1948, Theorem 8.2, P326]).
Let $\underset{\sim}{Q}, \underset{\sim}{R}, \underset{\sim}{S} \in \Sigma$ *and let* $Q_x \in \underset{\sim}{Q}$. *The order of the events* $[f^+_{RQ}(Q_x)]$, $[f^+_{SQ}(Q_x)]$ *on the right optical line through* $[Q_x]$ *is the same as the order of the instants*

$$\left(f_{QS} \circ f_{SQ}\right)^*(Q_x) \text{ and } \left(f_{QR} \circ f_{RQ}\right)^*(Q_x) \text{ in } \underset{\sim}{Q} .$$

This theorem is a consequence of the preceding definitions. It is used in the proof of Theorems 21 (§5.1), 22 (§5.2), 23 (§5.3), 24 (§5.3), 25 (§5.4), 26 (§5.5), 27 (§6.1), 28 (§6.1), 29 (§6.2), 30 (§6.3), Corollary 3 to 32 (§6.4), 33 (§6.4), 36 (§7.1), 37 (§7.2), the Corollary to 39 (§7.3), 40 (§7.3), 41 (§7.3), the Corollary to 41 (§7.3), 43 (§7.4), and 46 (§7.5).

PROOF. From the previous definitions. □

COROLLARY. *If* $\left(f_{QR} \circ f_{RQ}\right)^*(Q_x) = \left(f_{QS} \circ f_{SQ}\right)^*(Q_x)$,

then $\underset{\sim}{R}$ *and* $\underset{\sim}{S}$ *coincide at* $[f^+_{RQ}(Q_x)]$. □

§4.4 Betweenness Relation for *n* Particles

THEOREM 20. *Let* $\underset{\sim}{Q}, \underset{\sim}{R}, \underset{\sim}{S}, \underset{\sim}{T}$ *be distinct particles. Then*

(i) $\langle \underset{\sim}{Q}, \underset{\sim}{R}, \underset{\sim}{T} \rangle$ *and* $\langle \underset{\sim}{R}, \underset{\sim}{S}, \underset{\sim}{T} \rangle \Rightarrow \langle \underset{\sim}{Q}, \underset{\sim}{R}, \underset{\sim}{S}, \underset{\sim}{T} \rangle$, *and*

(ii) $\langle \underset{\sim}{Q}, \underset{\sim}{R}, \underset{\sim}{S} \rangle$ *and* $\langle \underset{\sim}{R}, \underset{\sim}{S}, \underset{\sim}{T} \rangle \Longrightarrow \langle \underset{\sim}{Q}, \underset{\sim}{R}, \underset{\sim}{S}, \underset{\sim}{T} \rangle$,

where $\langle \underset{\sim}{Q}, \underset{\sim}{R}, \underset{\sim}{S}, \underset{\sim}{T} \rangle$ *is a concise expression for the four statements:* $\langle \underset{\sim}{Q}, \underset{\sim}{R}, \underset{\sim}{S} \rangle$, $\langle \underset{\sim}{Q}, \underset{\sim}{R}, \underset{\sim}{T} \rangle$, $\langle \underset{\sim}{Q}, \underset{\sim}{S}, \underset{\sim}{T} \rangle$ *and* $\langle \underset{\sim}{R}, \underset{\sim}{S}, \underset{\sim}{T} \rangle$.

This theorem is a consequence of Axiom V (§2.7) and Theorem 3 (§2.7). It is used in the proof of Theorems 21 (§5.1), 22 (§5.2) and 24 (§5.3).

REMARK. At this stage we can not prove the proposition: "$\langle \underset{\sim}{Q}, \underset{\sim}{R}, \underset{\sim}{S} \rangle$ and $\langle \underset{\sim}{Q}, \underset{\sim}{R}, \underset{\sim}{T} \rangle \Rightarrow \langle \underset{\sim}{Q}, \underset{\sim}{R}, \underset{\sim}{S}, \underset{\sim}{T} \rangle$ or $\langle \underset{\sim}{Q}, \underset{\sim}{R}, \underset{\sim}{T}, \underset{\sim}{S} \rangle$", because $\underset{\sim}{S}$ and $\underset{\sim}{T}$ could "cross each other" at an ideal event and so Theorem 6 (§2.9) would not apply.

PROOF. Proposition (i) is a consequence of Theorem 3 (§2.7). Proposition (ii) is a consequence of the Axiom of Uniqueness of Extension of Optical Lines (Axiom V, §2.7). □

We shall also use the brackets < > to concisely represent betweenness relations for any number of particles; so, for example, $\langle \underset{\sim}{R}^{(1)}, \ldots \underset{\sim}{R}^{(n)}, \underset{\sim}{R}^{(n+1)}, \ldots \rangle$ means that, for all positive integers a, b, c with $0 \leqslant a \leqslant b \leqslant c$, $\langle \underset{\sim}{R}^{(a)}, \underset{\sim}{R}^{(b)}, \underset{\sim}{R}^{(c)} \rangle$. Similarly, we can extend the definition to apply to any linearly ordered indexing set such as the integers, the rationals, or the reals.

CHAPTER 5

COLLINEAR SUB-SPRAYS AFTER COINCIDENCE

In this chapter we will show that there are sub-SPRAYS which are "collinear after the event of coincidence" and which contain a "reflection of each particle in each other particle". In Theorem 25 (§5.4) we will show that the conditional completion of Chapter 3 is trivial, by showing that all instants are ordinary instants.

Since the Axiom of Compactness (Axiom XI, §2.13) applies to bounded sub-SPRAYs "after the event of coincidence", it is useful to modify some previous definitions so that they apply "after a certain event". Thus, the statement:
$\langle \underset{\sim}{Q}, \underset{\sim}{R}, \underset{\sim}{S} \rangle$ *after* $[R_c]$ means that, for all $R_x \in \underset{\sim}{R}$ with $R_x > R_c$, $|f^{-1}_{RQ}(R_x), R_x, f_{SR}(R_x)\rangle$ and $|f^{-1}_{RS}(R_x), R_x, f_{QR}(R_x)\rangle$.
There is a similar definition for the statement:
$[\underset{\sim}{Q}, \underset{\sim}{R}, \underset{\sim}{S}]$ *after* $[R_c]$. Both of these definitions can be extended to apply to any number of particles, as in the previous section.

§5.1 Collinearity of the Limit Particle

THEOREM 21 (Collinearity of Limit Particle)

Let $\underset{\sim}{Q}$ *be a particle with an instant* $Q_c \in \underset{\sim}{Q}$ *and let* $\{\underset{\sim}{R}^{(n)}: n=1,2,\cdots;\ \underset{\sim}{R}^{(n)} \in SPR[Q_c]\}$ *be a bounded sub-SPRAY of* $SPR[Q_c]$.

If $<\underset{\sim}{Q},\underset{\sim}{R}^{(1)},\cdots\underset{\sim}{R}^{(n)},\underset{\sim}{R}^{(n+1)}\cdots>$ *after* $[Q_c]$, *there is a unique particle* $\underset{\sim}{S} \in SPR[Q_c]$ *such that:*

(i) $<\underset{\sim}{Q},\cdots,\underset{\sim}{R}^{(n)},\underset{\sim}{R}^{(n+1)},\cdots\underset{\sim}{S}>$ *after* $[Q_c]$ *and*

(ii) *for any instant* $Q_x \in \underset{\sim}{Q}$ *with* $Q_x > Q_c$,

$$f_{QS} \circ f_{SQ}(Q_x) = \sup\left\{ f_{QR^{(n)}} \circ f_{R^{(n)}Q}(Q_x)\right\} \quad \text{and}$$

$$f^{-1}_{SQ} \circ f^{-1}_{QS}(Q_x) = \inf\left\{ f^{-1}_{R^{(n)}Q} \circ f^{-1}_{QR^{(n)}}(Q_x)\right\} .$$

We call $\underset{\sim}{S}$ *the limit particle of the sequence of particles* $(\underset{\sim}{R}^{(n)})$.

This theorem is a consequence of Axioms IV (§2.4), XI (§2.13) and Theorems 6 (§2.9), 19 (§4.3) and 20 (§4.4). It is used in the proof of Theorems 22 (§5.2), 23 (§5.3), 25 (§5.4) and 36 (§7.1).

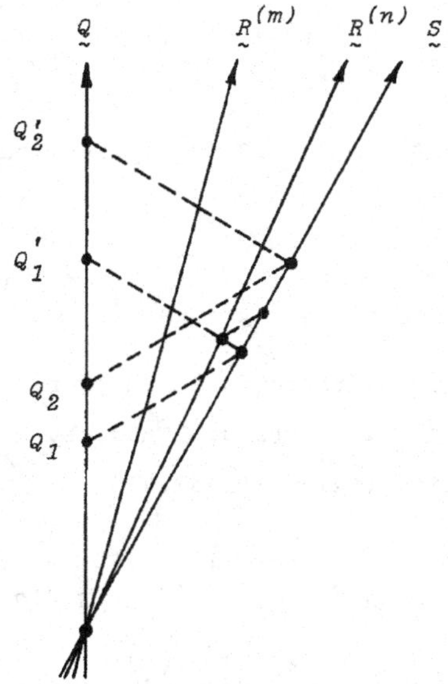

Fig. 10

PROOF (see Fig. 10).

(i) The set of particles $\{\underset{\sim}{R}^{(n)}: n=1,2,\cdots\}$ is an infinite bounded set and so by the Axiom of Compactness (Axiom XI, §2.13) it has a cluster particle $\underset{\sim}{S}$; that is, for any instants $Q_1, Q_2 \in \underset{\sim}{Q}$ with

$$Q_c < Q_1 < Q_2,$$

there is an infinite subset N of the positive integers such that, for all $n \in N$,

$$f_{SR^{(n)}} \circ f_{R^{(n)}S} \circ f_{SQ}(Q_1) < f_{SQ}(Q_2).$$

The Triangle Inequality (Axiom IV, §2.4) implies that, for all $n \in N$,

$$(1) \quad f_{SR^{(n)}} \circ f_{R^{(n)}Q}(Q_1) \leq f_{SR^{(n)}} \circ f_{R^{(n)}S} \circ f_{SQ}(Q_1) < f_{SQ}(Q_2).$$

Given any positive integer m, there is an integer $n \in N$ with $n \geq m$, and so $<\underset{\sim}{Q},\underset{\sim}{R}^{(m)},\underset{\sim}{R}^{(n)}>$ *after* $[Q_c]$. Thus, by collinearity and the Triangle Inequality (Axiom IV, §2.4),

$$\begin{aligned} f_{SR^{(m)}} \circ f_{R^{(m)}Q}(Q_1) &\leq f_{SR^{(n)}} \circ f_{R^{(n)}R^{(m)}} \circ f_{R^{(m)}Q}(Q_1) \\ &= f_{SR^{(n)}} \circ f_{R^{(n)}Q}(Q_1) \\ &< f_{SQ}(Q_2), \text{ by (1).} \end{aligned}$$

Now Q_2 was arbitrary apart from the requirement that $Q_2 > Q_1$, so for all positive integers m,

$$f_{SR^{(m)}} \circ f_{R^{(m)}Q}(Q_1) \leq f_{SQ}(Q_1).$$

By the Triangle Inequality and the previous equation, we see that for all positive integers m,

$$(2) \qquad f_{SR^{(m)}} \circ f_{R^{(m)}Q}(Q_1) = f_{SQ}(Q_1).$$

Now let $Q_1' \overset{def}{=} f_{QS} \circ f_{SQ}(Q_1)$, $Q_2' \overset{def}{=} f_{QS} \circ f_{SQ}(Q_2)$. The second of the inequalities (1), which applies for all $n \in N$, can now be written in the alternative form

$$(3) \qquad f_{QR^{(n)}} \circ f_{R^{(n)}S} \circ f^{-1}_{QS}(Q_1') \leqslant f_{QS} \circ f_{SR^{(n)}} \circ f_{R^{(n)}S} \circ f^{-1}_{QS}(Q_1') < Q_2'.$$

As before, for any positive integer m there is some integer $n \in N$ with $n \geqslant m$ and so by collinearity, the Triangle Inequality, and the above inequality,

$$\begin{aligned} f_{QR^{(m)}} \circ f_{R^{(m)}S} \circ f^{-1}_{QS}(Q_1') &\leqslant f_{QR^{(m)}} \circ f_{R^{(m)}R^{(n)}} \circ f_{R^{(n)}S} \circ f^{-1}_{QS}(Q_1') \\ &= f_{QR^{(n)}} \circ f_{R^{(n)}S} \circ f^{-1}_{QS}(Q_1') \\ &< Q_2' . \end{aligned}$$

Now Q_2' was arbitrary, apart from the requirement that $Q_2' > Q_1'$, so for all positive integers m,

$$f_{QR^{(m)}} \circ f_{R^{(m)}S} \circ f^{-1}_{QS}(Q_1') \leqslant Q_1'$$

and, by using the Triangle Inequality,

$$(4) \qquad f_{QR^{(m)}} \circ f_{R^{(m)}S} \circ f^{-1}_{QS}(Q_1') = Q_1' .$$

Equations (2) and (4) apply to any cluster particle $\underset{\sim}{S}$, for any instants $Q_1, Q_1' \in \underset{\sim}{Q}$ with $Q_1 > Q_c$ and $Q_1' > Q_c$, so for any cluster particle $\underset{\sim}{S}$ and for all positive integers m,

$$<\underset{\sim}{Q}, \underset{\sim}{R}^{(m)}, \underset{\sim}{S}> \textit{ after } [Q_c].$$

The proof of (i) is completed by applying Theorem 20(i) (§4.4).

(ii) Since $\left\{ f_{QR^{(n)}} \circ f_{R^{(n)}Q}(Q_1): n=1,2,\cdots \right\}$ is an increasing sequence, the Axiom of Compactness (Axiom XI, §2.3) and the Triangle Inequality (Axiom IV, §2.4) imply that for any instant $Q_0 \in \underset{\sim}{Q}$ with $Q_c < Q_0 < Q_1$, there is an integer M such that, for all $n > M$,

$$f_{QR^{(n)}} \circ f_{R^{(n)}Q}(Q_1) \geq f_{QS} \circ \left[f_{SR^{(n)}} \circ f_{R^{(n)}S} \right]^{-1} \circ f_{SQ}(Q_1)$$

$$> f_{QS} \circ f_{SQ}(Q_0).$$

Consequently

$$sup\left\{ f_{QR^{(n)}} \circ f_{R^{(n)}Q}(Q_1): n=1,2,\cdots \right\} \geq f_{QS} \circ f_{SQ}(Q_0),$$

and since Q_0 is arbitrary, subject to the condition $Q_0 < Q_1$, we see that

$$sup\left\{ f_{QR^{(n)}} \circ f_{R^{(n)}Q}(Q_1): n=1,2,\cdots \right\} \geq f_{QS} \circ f_{SQ}(Q_1).$$

Theorem 19 (§4.3), together with part (i) of this theorem, implies the opposite inequality, which establishes the first equation of (ii). In order toderive the second equation, let $Q_2 \in \underset{\sim}{Q}$ with $Q_0 < Q_1 < Q_2$. Then, by (i), there is an integer M such that, for all $n > M$,

$$f^{-1}_{R^{(n)}Q} \circ f^{-1}_{QR^{(n)}}(Q_1) \leq f^{-1}_{SQ} \circ \left[f_{SR^{(n)}} \circ f_{R^{(n)}S} \right]^{+1} \circ f^{-1}_{QS}(Q_1)$$

$$< f^{-1}_{SQ} \circ f^{-1}_{QS}(Q_2).$$

From this stage on, the proof is similar to the proof of the first equation.

By Theorem 19 (§4.3) and Theorem 6 (§2.9), there is only one (distinct) cluster particle $\underset{\sim}{S}$. □

§5.2 The Set of Intermediate Particles

We now demonstrate the existence of a sub-SPRAY which is collinear after the event of coincidence and "has no gaps".

THEOREM 22 (see Fig. 11)

Let $\underset{\sim}{Q}, \underset{\sim}{U}$ be distinct particles which coincide at the event $[Q_c]$ and let $Q_x > Q_c$.
Given any instant $\overline{Q}_y \in \overline{\underset{\sim}{Q}}$ with $Q_x < \overline{Q}_y < f_{QU} \circ f_{UQ}(Q_x)$,
there is a particle $\underset{\sim}{S} \in SPR[Q_c]$ such that:

(i) $f_{QS} \circ f_{SQ}(Q_x) = \overline{Q}_y = Q_y$ *and*

(ii) $\langle \underset{\sim}{Q}, \underset{\sim}{S}, \underset{\sim}{U} \rangle$ *after* $[Q_c]$.

REMARK. $\underset{\sim}{S} \in \Sigma^*$ which is a linearly ordered subspray after the event $[Q_c]$.

This theorem is a consequence of Axiom VI (§2.8) and Theorems 4 (§2.7), 19 (§4.3), 20 (§4.4) and 21 (§5.1). It is used in the proof of Theorems 23 (§5.3), 25 (§5.4), 26 (§5.5) and 57 (§9.1).

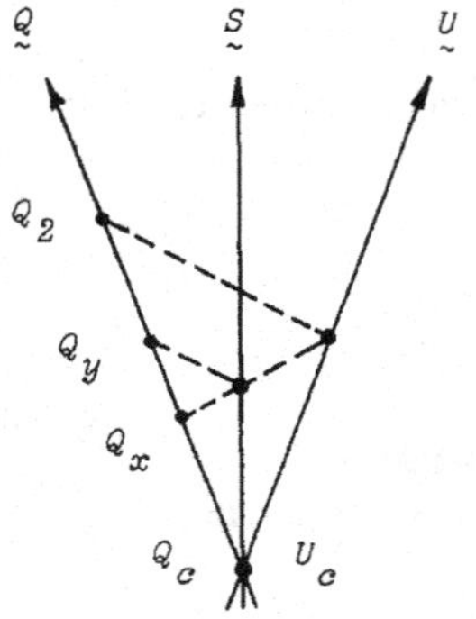

Fig. 11

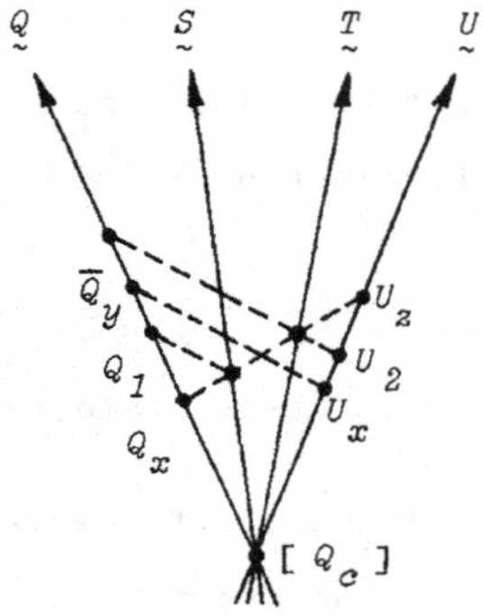

Fig. 12

PROOF (see Fig. 12)

By the Axiom of the Intermediate Particle (Axiom VI, §2.8), there is a particle $\underset{\sim}{W} \in SPR[Q_c]$ such that $\langle \underset{\sim}{Q}, \underset{\sim}{W}, \underset{\sim}{U} \rangle$ *after* $[Q_c]$, and $\underset{\sim}{Q} \neq \underset{\sim}{W} \neq \underset{\sim}{U}$. In this proof we shall consider particles from the set

$$\Sigma \overset{def}{=} \{\underset{\sim}{R}: \langle \underset{\sim}{Q}, \underset{\sim}{R}, \underset{\sim}{W}\rangle \textit{ after } [Q_c],\ \underset{\sim}{R} \in SPR[Q_c]\} \cup$$

$$\cup \{\underset{\sim}{R}: \langle \underset{\sim}{W}, \underset{\sim}{R}, \underset{\sim}{U}\rangle \textit{ after } [Q_c],\ \underset{\sim}{R} \in SPR[Q_c]\}$$

By Theorem 4 (§2.7) Σ is a collinear set of particles, and by Theorem 20 (§4.4), Σ is a partially ordered set. Thus, by the Hausdorff maximal principle, which is logically equivalent to the Axiom of Choice (see Rubin and Rubin [1963]), Σ has a maximal (with respect to set inclusion) linearly ordered subset Σ^*. By Theorem 19 (§4.3) and the previous theorem, there is a particle $\underset{\sim}{S} \in \Sigma^*$ such that

$$(1)\quad \underset{QS}{f} \circ \underset{SQ}{f}(Q_x) = sup\left\{\underset{QR}{f} \circ \underset{RQ}{f}(Q_x): \underset{QR}{f} \circ \underset{RQ}{f}(Q_x) \leqslant \overline{Q}_y,\ \underset{\sim}{R} \in \Sigma^*\right\}$$

$$\overset{def}{=} Q_1$$

$$\leqslant \overline{Q}_y.$$

Let $\Sigma' \overset{def}{=} \left\{\underset{\sim}{R}: \underset{QR}{f} \circ \underset{RQ}{f}(Q_x) \geqslant \overline{Q}_y,\ \underset{\sim}{R} \in \Sigma^*\right\}$;

let $\overline{U}_x \overset{def}{=} \underset{QU}{f^{-1}}(\overline{Q}_y)$ and let $U_z \overset{def}{=} \underset{UQ}{f}(Q_x)$.

For any particle $\underset{\sim}{R} \in \Sigma'$

$$\langle \underset{\sim}{Q}, \underset{\sim}{R}, \underset{\sim}{U}\rangle \textit{ after } [Q_c], \text{ so}$$

$$\underset{QU}{f} \circ \underset{RU}{f^{-1}} \circ \underset{UR}{f^{-1}} \circ \underset{UQ}{f}(Q_x) = \underset{QR}{f} \circ \underset{RQ}{f}(Q_x) \geqslant \overline{Q}_y ,$$

whence
$$\underset{RU}{f^{-1}} \circ \underset{UR}{f^{-1}}(U_z) \geqslant \overline{U}_x .$$

Let $\overline{U}_2 \overset{def}{=} inf\{\underset{RU}{f^{-1}} \circ \underset{UR}{f^{-1}}(U_z): \underset{\sim}{R} \in \Sigma'\} \geqslant \overline{U}_x$.

By Theorem 19 (§4.3) and the previous theorem, there is a particle $\underset{\sim}{T} \in \Sigma^*$ such that

$$f^{-1}_{TU} \circ f^{-1}_{UT}(U_z) = U_2 = \overline{U}_2 ,$$

and since $<\underset{\sim}{Q},\underset{\sim}{T},\underset{\sim}{U}>$ *after* $[Q_c]$,

$$f^{-1}_{TQ} \circ f^{-1}_{QT}(\overline{Q}_y) = f^{-1}_{UQ} \circ f_{UT} \circ f_{TU} \circ f^{-1}_{QU}(\overline{Q}_y)$$

$$= f^{-1}_{UQ} \circ f_{UT} \circ f_{TU}(\overline{U}_x)$$

$$\leqslant f^{-1}_{UQ} \circ f_{UT} \circ \ _{TU}(\overline{U}_2), \text{ so}$$

$$f^{-1}_{TQ} \circ f^{-1}_{QT}(\overline{Q}_y) \leqslant f^{-1}_{UQ}(U_z) = Q_x$$

and therefore

(2) $$f_{QT} \circ f_{TQ}(Q_x) \geqslant \overline{Q}_y .$$

By equations (1) and (2) together with Theorem 20 (§4.4), $<\underset{\sim}{Q},\underset{\sim}{S},\underset{\sim}{T},\underset{\sim}{U}>$ *after* $[Q_c]$. If $\underset{\sim}{S} \not\simeq \underset{\sim}{T}$, the Axiom of Existence of the Intermediate Particle (Axiom VI, §2.8) implies the existence of a particle between $\underset{\sim}{S}$ and $\underset{\sim}{T}$ and distinct from both; and by Theorem 19 (§4.3), this contradicts equation (1), equation (2), or the maximality of Σ^*. Thus $\underset{\sim}{S} \simeq \underset{\sim}{T}$, and equations (1) and (2) imply that

$$f_{QS} \circ f_{SQ}(Q_x) = Q_y = \overline{Q}_y ,$$

which completes the proof. □

CORALLARY 1 *Let $\underset{\sim}{Q},\underset{\sim}{U}$ be distinct particles which coincide at the event $[Q_c]$. Let $Q_o > Q_c$ and, for each integer n, let $Q_n \overset{def}{=} \left(f_{QU} \circ f_{UQ}\right)^n (Q_0)$. Then*

$$\bigcup_{n=-\infty}^{+\infty} [Q_n, Q_{n+1}] \subset \underset{\sim}{Q}, \text{ where } [Q_n, Q_{n+1}] = \{\overline{Q}_x : Q_n \leq \overline{Q}_x \leq Q_{n+1}, \overline{Q}_x \in \overline{Q}\} .$$

This corollary is a consequence of Theorem 8 (§3.2) and is used in the proofs of the next corollary and Theorem 25 (§5.4).

PROOF. By (i) of the preceding theorem, $[Q_0, Q_1] \subset \underset{\sim}{Q}$. Since (extended) signal functions are one-to-one, order-preserving, and map ordinary instants onto ordinary instants (Theorem 8 §3.2),

$$(f_{QR} \circ f_{RQ})^n : [Q_0, Q_1] \to [Q_n, Q_{n+1}] \subset \underset{\sim}{Q},$$

from which the stated set containment follows. □

The proof of the next theorem is based on a proof of the Intermediate Value Theorem of real variable theory (as in Fulks [1961]).

COROLLARY 2. *Let $\underset{\sim}{Q},\underset{\sim}{R},\underset{\sim}{S}$ be distinct particles which coincide at the event $[Q_c]$. Let $Q_o > Q_c$ and, for each integer n, let*

$$Q_n \overset{def}{=} (f_{QS} \circ f_{SQ})^n (Q_0).$$

If $f_{QR} \circ f_{RQ}(Q_0) < f_{QS} \circ f_{SQ}(Q_0)$,

then, for all $Q_x \in \bigcup_{n=-\infty}^{+\infty} [Q_n, Q_{n+1}]$,

$$f_{QR} \circ f_{RQ}(Q_x) < f_{QS} \circ f_{SQ}(Q_x) .$$

This corollary is a consequence of Theorems 5 (§2.9) and 9 (§3.2) and the previous corollary. It is used in the proofs of Theorems 23 (§5.3) and 24 (§5.3) and the Corollary to Theorem 25 (§5.4).

PROOF. We suppose the contrary; that is, we suppose there is an instant $Q_u \in \bigcup_{n=-\infty}^{+\infty} [Q_n, Q_{n+1}]$ such that

$$f_{QR} \circ f_{RQ}(Q_u) \geq f_{QS} \circ f_{SQ}(Q_u) ,$$

and deduce a contradiction.

Case 1. $Q_u < Q_0$ and $f_{QR} \circ f_{RQ} > f_{QS} \circ f_{SQ}(Q_u)$.
Let

$$\overline{Q}_w \overset{def}{=} sup\left\{Q_t\colon f_{QR} \circ f_{RQ}(Q_t) > f_{QS} \circ f_{SQ}(Q_t),\ Q_u \leq Q_t < Q_0,\ Q_t \in \underset{\sim}{Q}\right\},$$

and let

$$\overline{Q}_v \overset{def}{=} sup\left[\left\{Q_u\right\} \cup \left\{Q_t\colon f_{QR} \circ f_{RQ}(Q_t) < f_{QS} \circ f_{SQ}(Q_t),\right.\right.$$
$$\left.\left. Q_u < Q_t < \overline{Q}_w,\ Q_t \in \underset{\sim}{Q}\right\}\right] ;$$

whence by Theorem 9 (§3.2),

$$Q_u \leq \overline{Q}_v < \overline{Q}_w < Q_0 .$$

Then,

$$sup\left\{f_{QR} \circ f_{RQ}(Q_t)\colon \overline{Q}_v < Q_t < \overline{Q}_w,\ Q_t \in \underset{\sim}{Q}\right\} ,$$
$$\geq sup\left\{f_{QS} \circ f_{SQ}(Q_t)\colon \overline{Q}_v < Q_t < \overline{Q}_w,\ Q_t \in \underset{\sim}{Q}\right\}$$

and

$$\inf\left\{ f_{QR} \circ f_{RQ}(Q_t) : \overline{Q}_w < Q_t < Q_0,\ Q_t \in \underset{\sim}{Q} \right\}$$

$$\leqslant \inf\left\{ f_{QS} \circ f_{SQ}(Q_t) : \overline{Q}_w < Q_t < Q_0,\ Q_t \in \underset{\sim}{Q} \right\};$$

so by Theorem 9 (§3.2),

$f_{QR} \circ f_{RQ}(\overline{Q}_w) \geqslant f_{QS} \circ f_{SQ}(\overline{Q}_w)$ and $f_{QR} \circ f_{RQ}(\overline{Q}_w) \leqslant f_{QS} \circ f_{SQ}(\overline{Q}_w)$

respectively, whence

$$f_{QR} \circ f_{RQ}(\overline{Q}_w) = f_{QS} \circ f_{SQ}(\overline{Q}_w) .$$

By the previous corollary, $\overline{Q}_w$ is an ordinary instant and so by Theorem 5 (2.9),

$$f_{QR} \circ f_{RQ} = f_{QS} \circ f_{SQ} ,$$

which is a contradiction.

Case 2. $Q_u > Q_x$ and $f_{QR} \circ f_{RQ}(Q_u) > f_{QS} \circ f_{SQ}(Q_u)$.

The proof is similar to the proof of Case 1.

Case 3. $f_{QR} \circ f_{RQ}(Q_u) = f_{QS} \circ f_{SQ}(Q_u)$.

By Theorem 5 (§2.9),

$$f_{QR} \circ f_{RQ} = f_{QS} \circ f_{SQ} ,$$

which is a contradiction. □

§5.3 Mid-Way and Reflected Particles

Let $\underset{\sim}{Q}, \underset{\sim}{S}, \underset{\sim}{U}$ be particles which coincide at the event $[Q_c]$. If $<\underset{\sim}{Q}, \underset{\sim}{S}, \underset{\sim}{U}>$ after $[Q_c]$ and if

$$f_{SQ} \circ f_{QS} = f_{SU} \circ f_{US} ,$$

we say that $\underset{\sim}{S}$ is *mid-way between* $\underset{\sim}{Q}$ and $\underset{\sim}{U}$ (see Fig. 13). It transpires in §6.3 that

$$<\underset{\sim}{Q}, \underset{\sim}{S}, \underset{\sim}{U}> \textit{ after } [Q_c] \Longrightarrow <\underset{\sim}{Q}, \underset{\sim}{S}, \underset{\sim}{U}> .$$

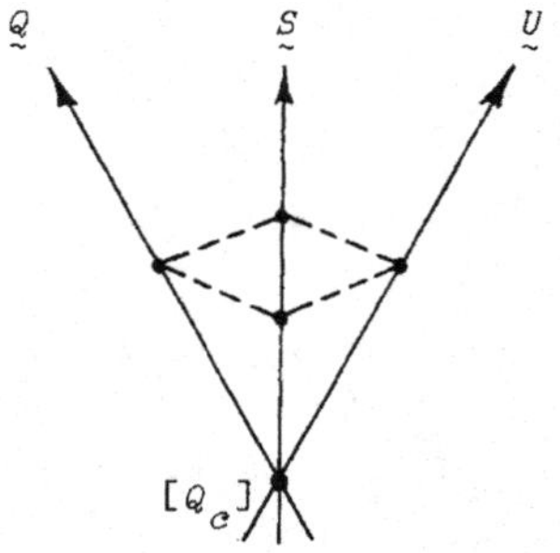

Fig. 13

THEOREM 23 (Existence of Mid-Way Particle)

Let $\underset{\sim}{Q}, \underset{\sim}{U}$ *be distinct particles which coincide at the event* $[Q_c]$ *Then there is a particle* $\underset{\sim}{S}$ *mid-way between* $\underset{\sim}{Q}$ *and* $\underset{\sim}{U}$.

This theorem is a consequence of Theorems 5 (§2.9), 15 (§3.7), 19 (§4.3), 21(§5.1), 22 (§5.2) and Corollary 2 of Theorem 22. It is used in the proof of Theorems 24 (§5.3), 37 (§7.2), 39 (§7.3), 44 (§7.4) and 47 (§7.5).

PROOF. As in the proof of the previous theorem, there exists a maximal (with respect to set inclusion) completely ordered collinear sub-SPRAY Σ^* which contains $\underset{\sim}{Q}$ and $\underset{\sim}{U}$ and particles between them. Let $Q_x \in \underset{\sim}{Q}$ be an instant such that $Q_x > Q_c$ and let $Q_z \overset{def}{=} f_{QU} \circ f_{UQ}(Q_x)$.

By Theorem 21 (§5.1) there is a particle $\underset{\sim}{S} \in \Sigma^*$ such that

$$(1)\quad f_{QS} \circ f_{SQ}(Q_x) = sup\left\{f_{QR} \circ f_{RQ}(Q_x) : \left(f_{QR} \circ f_{RQ}\right)^2(Q_x) \leq Q_z,\ \underset{\sim}{R} \in \Sigma^*\right\}$$

$$\overset{def}{=} Q_1 .$$

Case 1. (see Fig. 14)

Suppose $\left(f_{QS} \circ f_{SQ}\right)^2(Q_x) < Q_z$.

Then Theorem 15 (§3.7) and Theorem 22 (§5.2) imply the existence of an instant $Q_3 \in \underset{\sim}{Q}$ and a particle $\underset{\sim}{V} \in \Sigma^*$ such that

$$Q_1 < \left(f_{QV} \circ f_{VQ}\right)^{-1}(Q_z) = Q_3 < \left(f_{QS} \circ f_{SQ}\right)^{-1}(Q_z),$$

which means that $\langle \underset{\sim}{Q}, \underset{\sim}{S}, \underset{\sim}{V} \rangle$ *after* $[Q_c]$, since Σ^* is a linearly ordered subset of Σ. By Theorem 22 (§5.2) there is a particle $\underset{\sim}{T} \in \Sigma^*$ such that

$$f_{QT} \circ f_{TQ}(Q_x) = min\left\{Q_3,\ f_{QV} \circ f_{VQ}(Q_x)\right\} > Q_1$$

and, since $\underset{\sim}{Q}, \underset{\sim}{S}, \underset{\sim}{T}, \underset{\sim}{V} \in \Sigma^*$, which is linearly ordered,

$$\langle \underset{\sim}{Q}, \underset{\sim}{S}, \underset{\sim}{T}, \underset{\sim}{V} \rangle \text{ } after \text{ } [T_c] \text{ and } \underset{\sim}{S} \neq \underset{\sim}{T} .$$

But now, by Theorem 19 (§4.3),

$$\left(f_{QT} \circ f_{TQ}\right)^2 (Q_x) \leqslant f_{QT} \circ f_{TQ} (Q_3)$$

$$\leqslant f_{QV} \circ f_{VQ} (Q_3) = Q_z ,$$

which contradicts (1), and shows that the supposition is false.

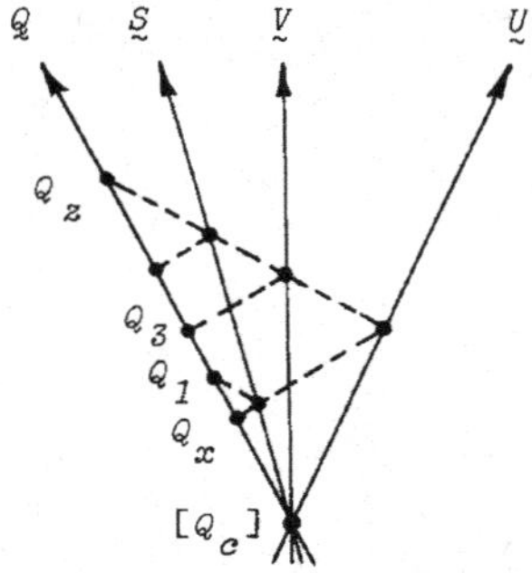

Fig. 14

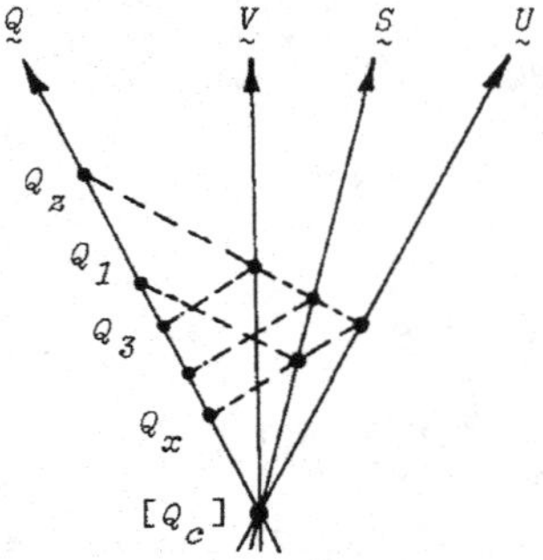

Fig. 15

Case 2. (see Fig. 15)

Suppose $\left(f_{QS} \circ f_{SQ}\right)^2 (Q_x) > Q_z$.

Then there is an instant $Q_3 \in \underset{\sim}{Q}$ and a particle $\underset{\sim}{V} \in \Sigma^*$ such that

$$Q_1 > \left(f_{QV} \circ f_{VQ}\right)^{-1} (Q_z) = Q_3 > \left(f_{QS} \circ f_{SQ}\right)^{-1} (Q_z)$$

which means that $\langle Q, V, S\rangle$ *after* $[Q_c]$. By Theorem 22 (§5.2) there is a particle $T \in \Sigma^*$ such that

$$f_{QT} \circ f_{TQ}(Q_x) = max\left\{Q_3, f_{QV} \circ f_{VQ}(Q_x)\right\} < Q_1$$

and $\langle Q, V, T, S\rangle$ *after* $[Q_c]$ and $T \neq S$. But now

$$\left(f_{QT} \circ f_{TQ}\right)^2(Q_x) \geqslant f_{QT} \circ f_{TQ}(Q_3)$$

$$\geqslant f_{QV} \circ f_{VQ}(Q_3) = Q_z ,$$

which contradicts (1) and shows that the supposition is false.

Having eliminated the previous two cases, we conclude that

$$\left(f_{QS} \circ f_{SQ}\right)^2(Q_x) = Q_z = f_{QU} \circ f_{UQ}(Q_x).$$

Also $\langle Q, S, U\rangle$ *after* $[Q_c]$, so letting $S_x \overset{def}{=} f_{SQ}(Q_x)$ and $S_z \overset{def}{=} f^{-1}_{QS}(Q_z)$,

$$f_{SQ} \circ f_{QS}(S_x) = S_z = f_{SU} \circ f_{US}(S_x) .$$

By Theorem 5 (§2.9)

$$f_{SQ} \circ f_{QS} = f_{SU} \circ f_{US} ,$$

which completes the proof. □

If a particle U is mid-way between two particles Q and W, we say that Q is a *reflection* of W in U and W is a reflection of Q in U. In the next theorem we show that for any two distinct particles $\underset{\sim}{Q}$ and $\underset{\sim}{U}$ which coincide at some event, there is at least one reflection of $\underset{\sim}{Q}$ in $\underset{\sim}{U}$, and all reflections of $\underset{\sim}{Q}$ in $\underset{\sim}{U}$ are permanently coincident; that is, there is a unique observer which we denote by the symbol $\hat{\underset{\sim}{Q}}_U$ (see §3.6). Before proving this existence theorem, we establish the following:

PROPOSITION (see Fig. 16)

Let $\underset{\sim}{Q}, \underset{\sim}{R}, \underset{\sim}{S}$ coincide at the event $[Q_c]$.

If

$$f_{RQ} \circ f_{QR} = f_{RS} \circ f_{SR} \quad \textit{and} \quad \left(f_{QR} \circ f_{RQ}\right)^2 = f_{QS} \circ f_{SQ},$$

then $\underset{\sim}{R}$ is mid-way between $\underset{\sim}{Q}$ and $\underset{\sim}{S}$.

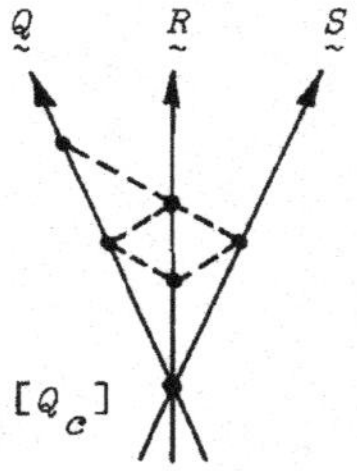

Fig. 16

PROOF. Since $f_{QR} \circ f_{RQ} \circ f_{QR} \circ f_{RQ} = f_{QS} \circ f_{SQ}$, it follows that

$$f_{QR} \circ f_{RS} \circ f_{SR} \circ f_{RQ} = f_{QS} \circ f_{SQ}.$$

By the Triangle Inequality (Axiom IV, §2.4),

$$f_{QR} \circ f_{RS} \geqslant f_{QS} \quad \text{and} \quad f_{SR} \circ f_{RQ} \geqslant f_{SQ}\text{, whence}$$

$$f_{SR} \circ f_{RQ} = f_{SQ} \quad \text{and} \quad f_{QR} \circ f_{RS} = f_{QS}\text{,}$$

which was to be proved. □

THEOREM 24 (Existence of Reflected Observer)

Let $\underset{\sim}{Q}, \underset{\sim}{U}$ *be distinct particles which coincide at some event* $[Q_c]$. *Then there is a unique observer* $\hat{\underset{\sim}{Q}}_U$.

This theorem is a consequence of Axiom VII (§2.9) and Theorems 19 (§4.3), 20 (§4.4), Corollary 2 to Theorem 22 (§5.2) and Theorem 24 (§5.3). It is used in the proof of Theorems 26 (§5.5), 36 (§7.1) and 42 (§7.3).

PROOF (see Fig. 17)

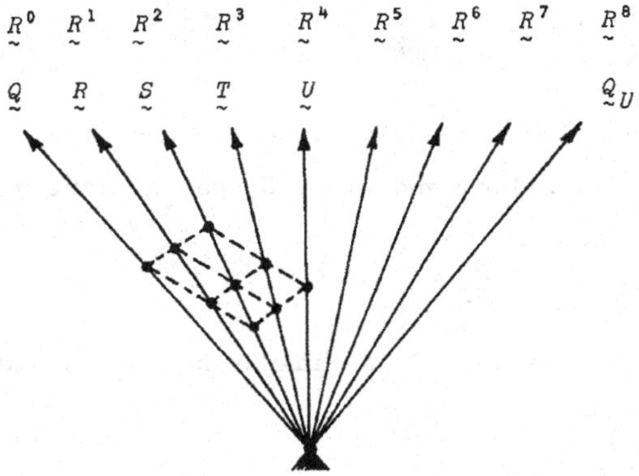

Fig. 17

Applying the previous theorem successively, there are particles:

$$(1)\quad \begin{cases} \underset{\sim}{S} \text{ such that } \underset{\sim}{S} \text{ is mid-way between } \underset{\sim}{Q} \text{ and } \underset{\sim}{U}; \\ \underset{\sim}{R} \text{ such that } \underset{\sim}{R} \text{ is mid-way between } \underset{\sim}{Q} \text{ and } \underset{\sim}{S}; \text{ and} \\ \underset{\sim}{T} \text{ such that } \underset{\sim}{T} \text{ is mid-way between } \underset{\sim}{S} \text{ and } \underset{\sim}{U}. \end{cases}$$

By Theorem 20 (§4.4),

$$(2)\qquad \langle \underset{\sim}{Q}, \underset{\sim}{R}, \underset{\sim}{S}, \underset{\sim}{T}, \underset{\sim}{U} \rangle \ \mathit{after}\ [Q_c].$$

We will now show that $\underset{\sim}{S}$ is mid-way between $\underset{\sim}{R}$ and $\underset{\sim}{T}$. If, for some instant $S_x \in \underset{\sim}{S}$ with $S_x > Q_c$,

$$(3)\qquad f_{SR} \circ f_{RS}(S_x) > f_{ST} \circ f_{TS}(S_x)$$

then, by Theorem 23 (§5.2),

$$\left(f_{SR} \circ f_{RS}\right)^2 (S_x) > \left(f_{ST} \circ f_{TS}\right)^2 (S_x).$$

The statements (1) imply that

$$\left(f_{SR} \circ f_{RS}\right)^2 = f_{SQ} \circ f_{QS} = f_{SU} \circ f_{US} = \left(f_{ST} \circ f_{TS}\right)^2$$

which shows that (3) is false. Similarly, we can show that the opposite inequality is false, and since S_x was arbitrary,

$$f_{SR} \circ f_{RS} = f_{ST} \circ f_{TS}\ .$$

This together with (2), shows that $\underset{\sim}{S}$ is mid-way between $\underset{\sim}{R}$ and $\underset{\sim}{T}$.

In order to simplify the proof, we shall now substitute the particle symbols $\underset{\sim}{R}^0, \underset{\sim}{R}^1, \underset{\sim}{R}^2, \underset{\sim}{R}^3, \underset{\sim}{R}^4$ for $\underset{\sim}{Q}, \underset{\sim}{R}, \underset{\sim}{S}, \underset{\sim}{T}, \underset{\sim}{U}$ respectively and f_{nm} will be substituted for $f_{R^n R^m}$. We now let $P(m)$ represent the proposition:

(4) *For* $n=1, 2, \cdots m$,

(i) $\langle R^0, R^1, \cdots\cdots, R^m, R^{m+1}\rangle$ *after* $[Q_c]$

(ii) $f_{n(n-1)} \circ f_{(n-1)n} = f_{n(n+1)} \circ f_{(n+1)n}$ *and*

(iii) *for all* $R_x^{n-1} > Q_c$,

$$\left[f_{(n-1)n} \circ f_{n(n-1)}\right]^2 \left[R_x^{n-1}\right] = \left[f_{(n-1)(n+1)} \circ f_{(n+1)(n-1)}\right]\left[R_x^{n-1}\right].$$

The result of the preceding paragraph, together with (1) and (2) can be summarised as:

(5) $P(3)$ is true.

The Axiom of Isotropy of SPRAYs (Axiom VII, §2.9) implies that, for any isotropy mapping ϕ, if $P(m)$ is true, then $\phi(P(m))$ is true where $\phi(P(m))$ is the statement:

(6) Let $f_{\phi(j)\phi(k)} \overset{def}{=} f_{\phi(R^j)\phi(R^k)}$. For $n=1, 2, \cdots, m$,

(i) $\langle \phi(\underset{\sim}{R}^0), \phi(\underset{\sim}{R}^1), \cdots \phi(\underset{\sim}{R}^{m+1})\rangle$ *after* $[Q_c]$,

(ii) $f_{\phi(n)\phi(n-1)} \circ f_{\phi(n-1)\phi(n)} = f_{\phi(n)\phi(n+1)} \circ f_{\phi(n+1)\phi(n)}$ and

(iii) for all $\phi(R_x^{n-1}) > R_c$,

$$\left[f_{\phi(n-1)\phi(n)} \circ f_{\phi(n)\phi(n-1)}\right]^2 \left[\phi(R_x^{n-1})\right] =$$

$$= \left[f_{\phi(n-1)\phi(n+1)} \circ f_{\phi(n+1)\phi(n-1)}\right]\left[\phi(R_x^{n-1})\right].$$

Since $P(3)$ is true, the Axiom of Isotropy of SPRAYs (Axiom VII, §2.9) and the first of equations (4), with $n = 3$, imply that there is an isotropy mapping ϕ such that

$$(7) \qquad \phi(\underset{\sim}{R}^3) \simeq \underset{\sim}{R}^3 \text{ and } \phi(\underset{\sim}{R}^2) \simeq \underset{\sim}{R}^4.$$

By the ordering relations of (4) and (6), together with Theorem 20 (§4.4) and Theorem 19 (§4.3),

$$(8) \quad \langle \underset{\sim}{R}^0, \underset{\sim}{R}^1, \underset{\sim}{R}^2 \simeq \phi(\underset{\sim}{R}^4), \underset{\sim}{R}^3 \simeq \phi(\underset{\sim}{R}^3), \underset{\sim}{R}^4 \simeq \phi(\underset{\sim}{R}^2), \phi(\underset{\sim}{R}^1), \phi(\underset{\sim}{R}^0) \rangle \quad after\ [Q_c].$$

Let $\underset{\sim}{R}^{6-n} \overset{def}{=} \phi(\underset{\sim}{R}^n)$, for $n=0,1$. We observe that $\phi\circ\phi(\underset{\sim}{R}^1) \simeq \underset{\sim}{R}^1$ and $\phi\circ\phi(\underset{\sim}{R}^0) \simeq \underset{\sim}{R}^0$, so

$$(9) \qquad \underset{\sim}{R}^{6-n} \simeq \phi(\underset{\sim}{R}^n),\ for\ n=0,1,\cdots 6.$$

By (4), (5), (6), (8) and (9) it follows that:

(10) $P(5)$ is true.

By a procedure similar to that of the preceding paragraph, we can show that

$P(9)$ is true.

Now by a few successive applications of equations (4),

$$f_{40} \circ f_{04} = f_{48} \circ f_{84} \quad (after\ [Q_c]).$$

That is, $\underset{\sim}{R}^8$ is a reflection of $\underset{\sim}{R}^0$ in $\underset{\sim}{R}^4$, so $\underset{\sim}{R}^8$ is a reflection of $\underset{\sim}{Q}$ in $\underset{\sim}{U}$. By Theorem 19 (§4.3), $\hat{\underset{\sim}{Q}}_U$ is unique. This completes the proof. □

COROLLARY (see Fig. 18)

Let $\underset{\sim}{S}^0, \underset{\sim}{S}^1$ *be distinct particles which coincide at the event* $[S_c^0]$. *There is a sub-SPRAY* $\{\underset{\sim}{S}^n : n=0,\pm 1,\pm 2,\ldots;\ \underset{\sim}{S}^n \in SPR[S_c^0]\}$ *such that, for all integers m, n and for any instant* $S_x^m \in \underset{\sim}{S}^m$ *with* $S_x^m > S_c^0$,

$$\left(f_{mn} \circ f_{nm}\right)^*(S_x^m) = \left(f_{m(m+1)} \circ f_{(m+1)m}\right)^{n-m}(S_x^m),$$

where $f_{mn} \overset{\text{def}}{=} f_{S^m S^n}$.

This corollary is used in the proofs of Theorem 25 (§5.4), the Corollary to Theorem 39 (§7.3) and Theorem 57 (§9.1).

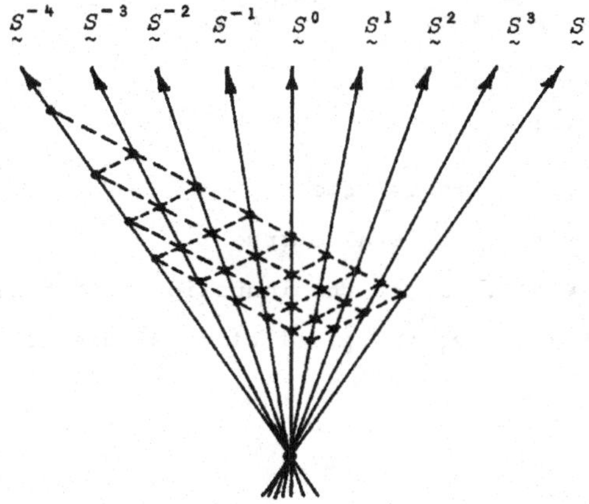

Fig. 18

PROOF (by induction)

For $n > 1$, let $S^{n+1} \in \hat{S}^{n+1}$ where $\hat{S}^{n+1}$ is defined as the unique reflection of $\hat{S}^{n-1}$ in $\hat{S}^{n}$.

For $n \leqslant 0$, let $S^{n-1} \in \hat{S}^{n-1}$ where $\hat{S}^{n-1}$ is defined as the unique reflection of $\hat{S}^{n+1}$ in $\hat{S}^{n}$.

Case 1. $m < n$

As an induction hypothesis, suppose that for all $S_x^k \in S^k$ with $S_x^k > S_c^0$ and for all k,ℓ with $1 \leqslant \ell - k \leqslant n$,

$$f_{k\ell} \circ f_{\ell k}(S_x^k) = \left(f_{k(k+1)} \circ f_{(k+1)k}\right)^{\ell-k}(S_x^k)\,.$$

Then, since $\langle S^k, S^{k+1}, S^\ell\rangle$ *after* $[S_c^0]$,

$$f_{k(\ell+1)} \circ f_{(\ell+1)k}(S_x^k) = f_{k(k+1)} \circ f_{(k+1)(\ell+1)} \circ f_{(\ell+1)(k+1)} \circ f_{(k+1)k}(S_x^k)$$

$$= f_{k(k+1)} \circ \left(f_{(k+1)(k+2)} \circ f_{(k+2)(k+1)}\right)^{\ell-k} \circ f_{(k+1)k}(S_x^k)$$

$$= f_{k(k+1)} \circ \left(f_{(k+1)k} \circ f_{k(k+1)}\right)^{\ell-k} \circ f_{(k+1)k}(S_x^k)$$

$$= \left(f_{k(k+1)} \circ f_{(k+1)k}\right)^{\ell+1-k}(S_x^k).$$

So if the induction hypothesis is true for all k,ℓ with $\ell - k \leqslant n$, it is also true for all k,ℓ with $\ell-k \leqslant n+1$. Since the induction hypothesis is (trivially) true for $n = 1$, the proof of this case is complete.

Case 2. $m > n$

In accordance with the definition of modified record functions, the proof is analogous to that of Case 1 but with inverse functions instead of functions.

Case 3. $m = n$

This trivial case completes the proof. □

§5.4 All Instants Are Ordinary Instants.

THEOREM 25. *All instants are ordinary instants.*

This theorem is a consequence of Axioms VIII (§2.10), IX (§2.11) and X (§2.12) and Theorems 6 (§2.9), 11 (§3.4), 12 (§3.5), 19 (§4.3), 21 (§5.1), 22 (§5.2), Corollary 1,2 of Theorem 22 (§5.2) and the Corollary to Theorem 24 (§5.3). It is used in the proof of the Corollary to Theorem 25 (§5.4), Theorem 26 (§5.5) and the Corollary to Theorem 40 (§7.3).

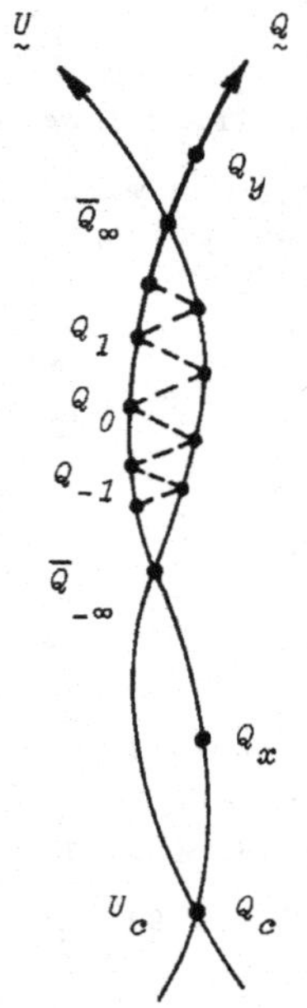

Fig. 19

PROOF (see Fig. 18)

Theorem 11 (§3.4) implies the existence of instants (of $\underset{\sim}{Q}$) before and after Q_0. Let Q_c be an instant of $\underset{\sim}{Q}$ with $Q_c < Q_0$. By the Axiom of Incidence (Axiom IX, §2.11), the Axiom of Connectedness (Axiom X, §2.12) and the Axiom of Dimension (Axiom VIII, §2.10), there is a particle $\underset{\sim}{U} \neq \underset{\sim}{Q}$ such that $\underset{\sim}{U}$ coincides with $\underset{\sim}{Q}$ at $[Q_c]$.

For each integer n, let $Q_n \stackrel{def}{=} \left(f_{QU} \circ f_{UQ}\right)^n (Q_0)$. By Corollary 1 of Theorem 22 (§5.2) and Theorem 12 (§3.5),

(1) $$\bigcup_{n=-\infty}^{+\infty} [Q_n, Q_{n+1}] \subset \{Q_x\colon Q_x > Q_c, Q_x \in \underset{\sim}{Q}\}$$

We now show that these two subsets are equal, by excluding two exceptional cases.

Case 1. Suppose there is an ordinary instant $Q_y \in \underset{\sim}{Q}$ such that, for all integers n, $Q_y > Q_n$. By the corollary to the previous theorem, there is a sub-SPRAY $\{\underset{\sim}{U}^n\colon n=0,1,2,\cdots;\ \underset{\sim}{U}^n \in SPRAY[Q_c]\}$ which is collinear after $[Q_c]$, such that $\underset{\sim}{Q} \simeq \underset{\sim}{U}^0$, $\underset{\sim}{U} \simeq \underset{\sim}{U}^1$ and for which $\overline{Q}_\infty \stackrel{def}{=} \sup_n Q_n = \sup_n\left\{f_{on} \circ f_{no} (Q_0)\right\} < Q_y$.

By Theorem 21 (§5.1), there is a limit particle $\underset{\sim}{S} \in SPRAY[Q_c]$ such that, for all positive integers n,

(2) $<\underset{\sim}{Q}\simeq\underset{\sim}{U}^0, \cdots, \underset{\sim}{U}^n, \underset{\sim}{U}^{n+1}, \cdots, \underset{\sim}{S}>$ *after* $[Q_c]$ and $\overline{Q}_\infty = Q_\infty = f_{QS} \circ f_{SQ} (Q_0)$.

Also,

$$\overline{Q}_\infty = \sup Q_n = \sup Q_{n+1} = \sup\left\{f_{on} \circ f_{no} (Q_1)\right\} < Q_y$$

and as before, there is a limit particle $\underset{\sim}{T} \in SPRAY[Q_c]$ such that, for all positive integers n,

(3) $<\underset{\sim}{Q}\simeq\underset{\sim}{U}^0, \cdots, \underset{\sim}{U}^n, \underset{\sim}{U}^{n+1}, \cdots, \underset{\sim}{T}>$ *after* $[Q_c]$ and $\overline{Q}_\infty = Q_\infty = f_{QT} \circ f_{TQ} (Q_1)$.

By Corollary 1 of Theorem 22,

$$\bigcup_{n=-\infty}^{+\infty} [Q_n, Q_{n+1}] \subseteq \underset{\sim}{Q} ,$$

so Theorem 6(§2.9), Theorem 19 (§4.3) and Corollary 2 of Theorem 22 (§5.2) imply that

$$<\underset{\sim}{Q},\underset{\sim}{T},\underset{\sim}{S}> \; after[inf\{Q_n\}] \text{ and } before[Q_\infty] \text{ , and}$$

$$\underset{\sim}{T} \nparallel \underset{\sim}{S} \text{ ,}$$

which contradicts (2). Hence there is no ordinary instant Q_y such that, for all integers n , $Q_y > Q_n$.

Case 2. Suppose there is an ordinary instant $Q_x \in \underset{\sim}{Q}$ with $Q_x > Q_c$, such that for all integers n, $Q_x < Q_n$. By Theorem 6 (§2.9), $\underset{\sim}{U}$ does not coincide with $\underset{\sim}{Q}$ at $[Q_x]$. By Theorem 12 (§3.5), for all negative integers n,

$$Q_c < Q_x < Q_n = \left(f_{QU} \circ f_{UQ}\right)^n Q_0 < Q_0 \in \underset{\sim}{Q}$$

As in Case 1, there is a set of particles $\{\underset{\sim}{U}^n: n=0,1,\cdots\}$ such that

$$\overline{Q}_{-\infty} \overset{def}{=} \inf_n\left\{\left(f_{on} \circ f_{no}\right)^{-1} Q_0\right\} > Q_x .$$

From this stage on, the proof is similar to the proof for Case 1; so there is no ordinary instant $Q_x \in \underset{\sim}{Q}$ such that, for all integers n, $Q_c < Q_x < Q_n$; whence $\overline{Q}_{-\infty} = Q_c$.

We have now shown that

$$(4) \qquad \bigcup_{n=0}^{\infty} [Q_c, Q_n] = \{Q_x \colon Q_x \geqslant Q_c, Q_x \in \underset{\sim}{Q}\}.$$

But Q_c was arbitrary, and since for any $\overline{Q}_w \in \underset{\sim}{\overline{Q}}$ there is some (ordinary) instant $Q_v \in \underset{\sim}{Q}$ such that $Q_v < \overline{Q}_w$, it follows by taking $Q_c = Q_v$, that $\overline{Q}_w \in \underset{\sim}{Q}$. Thus $\underset{\sim}{\overline{Q}} \subseteq \underset{\sim}{Q}$, which completes the proof. □

An immediate consequence of (the proof of) this theorem is the following stronger version of Corollary 2 of Theorem 22 (§5.2):

COROLLARY. *Let $\underset{\sim}{Q}, \underset{\sim}{R}, \underset{\sim}{S}$ be distinct particles which coincide at some event $[Q_c]$. If, for some instant $Q_o \in \underset{\sim}{Q}$ with $Q_o > Q_c$,*

$$f_{QR} \circ f_{RQ}(Q_0) < f_{QS} \circ f_{SQ}(Q_0)$$

then, for all $Q_x \in \underset{\sim}{Q}$ with $Q_x > Q_c$,

$$f_{QR} \circ f_{RQ}(Q_x) < f_{QS} \circ f_{SQ}(Q_x).$$

This corollary is a consequence of Corollary 2 of Theorem 22, and it is used in the proof of Theorem 26 (§5.5).

PROOF. By Corollary 2 to Theorem 22 (§5.2). □

§5.5 Properties of Collinear Sub-SPRAYs After Coincidence

Given two distinct particles $\underset{\sim}{Q}, \underset{\sim}{T}$ which coincide at some event $[Q_c]$, we define the *collinear sub-SPRAY*

$$CSP \langle \underset{\sim}{Q}, \underset{\sim}{T} \rangle \stackrel{def}{=} \{\underset{\sim}{R}: [\underset{\sim}{Q}, \underset{\sim}{T}, \underset{\sim}{R}] \text{ after } [Q_c],\ \underset{\sim}{R} \in SPR[Q_c]\},$$

and

$$csp \langle \underset{\sim}{Q}, \underset{\sim}{T} \rangle \stackrel{def}{=} \{[R_x]:\ R_x > Q_c,\ \underset{\sim}{R} \in CSP\langle \underset{\sim}{Q}, \underset{\sim}{T} \rangle\}.$$

Note that a *collinear sub-SPRAY (csp)* is a set of events, whereas a spray (*spr*) is a set of instants. In a *1+1* - dimensional Minkowski space-time, a collinear sub-spray is "a set of events contained within the upper half of a light cone".

Some of the previous results can now be stated concisely in the following:

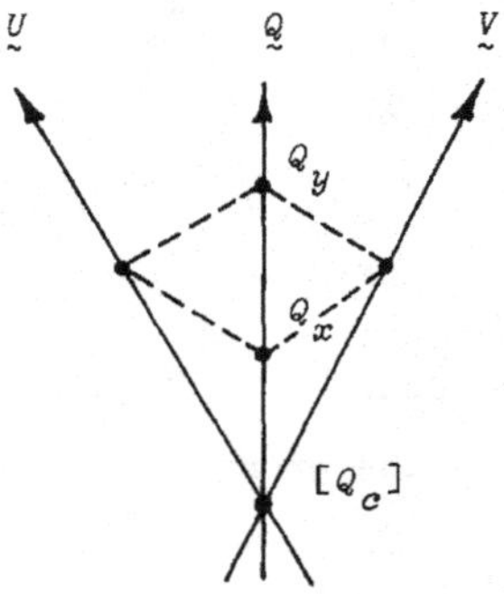

Fig. 20

THEOREM 26 (Collinear subSPRAYs after coincidence)

Let $\underset{\sim}{Q}, \underset{\sim}{R}$ *and* $\underset{\sim}{S}, \underset{\sim}{T}$ *be pairs of distinct particles which coincide at the event* $[Q_c]$.
Then

(i) *CSP* $<\underset{\sim}{Q}, \underset{\sim}{R}>$ *is a set of particles which is collinear after* $[Q_c]$ *and*

$$CSP<\underset{\sim}{Q},\underset{\sim}{R}> = \{\underset{\sim}{S}: <\underset{\sim}{S},\underset{\sim}{Q},\underset{\sim}{R}> \text{ after } [Q_c], \; <\underset{\sim}{Q},\underset{\sim}{S},\underset{\sim}{R}> \text{ after } [Q_c], \text{ or } <\underset{\sim}{Q},\underset{\sim}{R},\underset{\sim}{S}> \text{ after } [Q_c];\; \underset{\sim}{S} \in SPR[Q_c]\},$$

(ii) $\underset{\sim}{S}, \underset{\sim}{T} \in CSP<\underset{\sim}{Q},\underset{\sim}{R}> \Rightarrow CSP<\underset{\sim}{Q},\underset{\sim}{R}> = CSP<\underset{\sim}{S},\underset{\sim}{T}>$,

(iii) $\underset{\sim}{S}, \underset{\sim}{T} \in CSP<\underset{\sim}{Q},\underset{\sim}{R}> \Rightarrow \hat{S}_{\underset{\sim}{T}}, \hat{T}_{\underset{\sim}{S}} \subset CSP<\underset{\sim}{Q},\underset{\sim}{R}>$,

(iv) *the relation "to the right of" between particles in* $CSP<\underset{\sim}{Q},\underset{\sim}{R}>$ *after* $[Q_c]$ *is a linear ordering, and*

(v) *for any instants* $Q_x, Q_y \in \underset{\sim}{Q}$ *with* $Q_c < Q_x < Q_y$, *there are particles* $\underset{\sim}{U}, \underset{\sim}{V} \in CSP<\underset{\sim}{Q},\underset{\sim}{R}>$ *such that* $<\underset{\sim}{U},\underset{\sim}{Q},\underset{\sim}{R}>$ *after* $[Q_c]$ *and* $<\underset{\sim}{U},\underset{\sim}{Q},\underset{\sim}{V}>$ *after* $[Q_c]$ *and*

$$f_{QU} \circ f_{UQ}(Q_x) = f_{QV} \circ f_{VQ}(Q_x) = Q_y \quad \textit{(see Fig. 20)}.$$

This theorem is a consequence of Theorems 4 (§2.7), 17 (§4.2), 19 (§4.3), 22 (§5.2), 24 (§5.3), 25 (§5.4), and its corollary. It is used in the proof of Theorems 27 (§6.1), 28 (§6.1), 29 (§6.2), 30 (§6.3), 31 (§6.3), Corollary 3 to Theorem 32 (§6.4), and Theorem 33 (§6.4).

PROOF. (i) By Theorem 24 (§5.3) there is an observer
$\hat{\underset{\sim}{Q}}_R \subset CSP <Q,R>$ such that $<\hat{\underset{\sim}{Q}}_R, \hat{\underset{\sim}{Q}}_{\underset{\sim}{R}}>$ *after* $[Q_c]$.
By Theorem 4 (§2.7) optical lines containing instants (after
Q_c) from both $\underset{\sim}{Q}$ and $\underset{\sim}{R}$, are uniquely characterised. This
proves the first proposition; the second proposition is a
consequence of the previous corollary, Theorem 19 (§4.3)
and Theorem 17 (§4.2).

(ii) By (i), $CSP <Q,R>$ is a collinear set of particles
after $[Q_c]$, and is uniquely specified by any two distinct
particles contained in it.

(iii) As with (ii), this is a consequence of (i).

(iv) As with the second proposition of (i), this is a consequence of the previous corollary and Theorem 19 (§4.3).

(v) As in the proof of Theorem 25 (§5.4) (see equation (4)),
if we let $R^0 \in \hat{\underset{\sim}{Q}}$ and $R^1 \in \hat{\underset{\sim}{R}}$, there is some $\underset{\sim}{R}^n \in CSP <R^0,R^1>$
such that

$$\underset{on}{f} \circ \underset{no}{f} [Q_x] > [Q_y].$$

Now Theorem 22 (§5.2) implies the existence of $\underset{\sim}{V}$ such that
$<\underset{\sim}{R}^0, \underset{\sim}{V}, R^n>$ *after* $[Q_c]$ and

$$\underset{QV}{f} \circ \underset{VQ}{f} (Q_x) = Q_y.$$

Now take any particle $\underset{\sim}{U} \in \hat{\underset{\sim}{V}}_Q$. □

CHAPTER 6

COLLINEAR PARTICLES

In this chapter we show that there are collinear sets of particles which have many properties analogous to those of coplanar sets of lines in absolute geometry. These properties are described more fully in §6.4.

§6.1 Basic Theorems

The main result of this chapter is contained in the next theorem. The proof of some theorems would have been simplified if the Axiom of Compactness (Axiom XI, §2.13) had been stated so as to apply "before coincidence" as well as "after coincidence": the Second Collinearity Theorem (Theorem 30, §6.3) would be a trivial extension of the First Collinearity Theorem (Theorem 27, §6.1); similarly Theorem 31 (§6.4) would be a trivial extension of Theorem 26 (§5.5).

THEOREM 27 (First Collinearity Theorem)

Let $\underset{\sim}{Q}, \underset{\sim}{R}, \underset{\sim}{S}$ *be (distinct) particles with instants* $Q_a, Q_b \in \underset{\sim}{Q}$; $R_b, R_c \in \underset{\sim}{R}$; $S_a, S_c \in \underset{\sim}{S}$ *(see Fig. 21).*

If $S_a \simeq Q_a < Q_b \simeq R_b < R_c \simeq S_c$, *then*

$csp\langle \underset{\sim}{R}, \underset{\sim}{S} \rangle \subset csp\langle \underset{\sim}{Q}, \underset{\sim}{R} \rangle \subset csp\langle \underset{\sim}{Q}, \underset{\sim}{S} \rangle$.

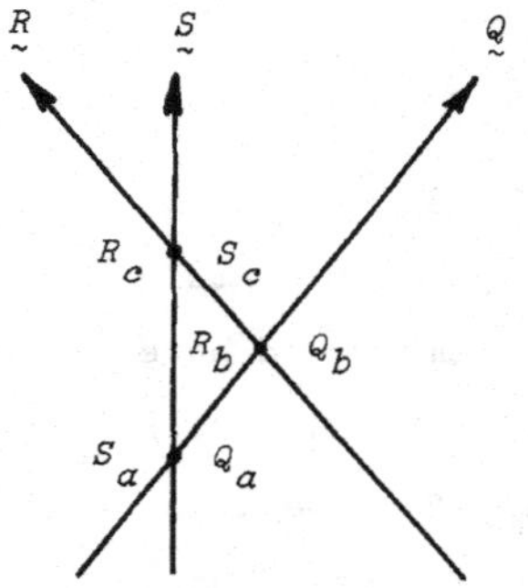

Fig. 21

This theorem is a consequence of Theorems 4(§2.7), 19 (§4.3) and 26 (§5.5). It is used in the proof of Theorems 28 (§6.1), 29 (§6.2), 30 (§6.3) and Corollary 1 to 30 (§6.3), and Theorem 32 (§6.4).

PROOF. By the preceding theorem and Theorem 4 (§2.7), any optical line containing two events from a collinear sub-spray (after the event of coincidence) is uniquely determined. In particular, since $R_c \simeq S_c$,

$$[R_c] = [S_c] \in csp\langle \underset{\sim}{Q}, \underset{\sim}{R}\rangle \cap csp\langle \underset{\sim}{Q}, \underset{\sim}{S}\rangle.$$

Therefore:

(1) any event which is in optical line with either the events $[R_c]$ and $[f_{QR}(R_c)]$, or with the events $[R_c]$ and $[f^{-1}_{RQ}(R_c)]$, is contained in both $csp\langle \underset{\sim}{Q}, \underset{\sim}{R}\rangle$ *and* $csp\langle \underset{\sim}{Q}, \underset{\sim}{S}\rangle$.

We now choose any particle $\underset{\sim}{T} \in CSP\langle \underset{\sim}{Q}, \underset{\sim}{R}\rangle$ *such that* $\langle \underset{\sim}{R}, \underset{\sim}{Q}, \underset{\sim}{T}\rangle$ *after* $[Q_b]$ and $\underset{\sim}{Q} \neq \underset{\sim}{T}$. By the previous theorem and Theorem 4 (§2.7),

(2) If we can show that

$$\{[T_w]: T_w > Q_b, T_w \in \underset{\sim}{T}\} \subset csp\langle \underset{\sim}{Q}, \underset{\sim}{S}\rangle,$$

it will follow that

$$csp\langle \underset{\sim}{Q}, \underset{\sim}{R}\rangle = csp\langle \underset{\sim}{Q}, \underset{\sim}{T}\rangle \subset csp\langle \underset{\sim}{Q}, \underset{\sim}{S}\rangle.$$

Let $T_1 \overset{def}{=} f^{-1}_{RT}(R_c)$, $T_2 \overset{def}{=} f_{TR}(R_c)$.

By proposition (1) and the definition of $\underset{\sim}{T}$,

(3) $$[T_1], [T_2] \in csp\langle \underset{\sim}{Q}, \underset{\sim}{S}\rangle.$$

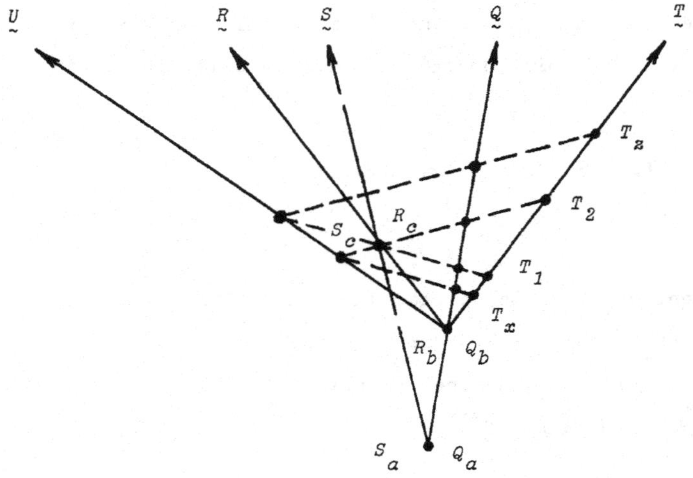

Fig. 22 In this diagram, the same particle $\underset{\sim}{U}$ is depicted for both cases 1 and 2. In general, the particle $\underset{\sim}{U}$ depends on the particular instants T_z and T_x, respectively.

Case 1. $T_w > T_2$ (see Fig. 22)

Let $T_w = T_z > T_2$. By the previous theorem, there is a particle $\underset{\sim}{U} \in CSP\langle\underset{\sim}{Q},\underset{\sim}{R}\rangle$ such that $\langle\underset{\sim}{U},\underset{\sim}{R},\underset{\sim}{Q},\underset{\sim}{T}\rangle$ after $[Q_b]$ and

$$f_{TU} \circ f_{UT}(T_1) = T_z \text{ , so}$$

$$f_{TQ} \circ f_{QU} \circ f_{UR} \circ f_{RQ} \circ f_{QT}(T_1) = T_z \text{ .}$$

Therefore

$$|T_1, f^{-1}_{RQ}(R_c), R_c, f_{UR} \circ f_{RQ} \circ f_{QT}(T_1)>$$

and by proposition (1)

$$[f_{UR} \circ f_{RQ} \circ f_{QT}(T_1)] \in csp<\underset{\sim}{Q},\underset{\sim}{S}>.$$

Also, trivially

$$[f_{QU} \circ f_{UR} \circ f_{RQ} \circ f_{QT}(T_1)] \in csp<\underset{\sim}{Q},\underset{\sim}{S}>.$$

Since $[T_z]$ is in optical line with the distinct events

$$[f_{UR} \circ f_{RQ} \circ f_{QT}(T_1)] \text{ and } [f_{QU} \circ f_{UR} \circ f_{RQ} \circ f_{QT}(T_1)],$$

it follows that

$$[T_z] \in csp<\underset{\sim}{Q},\underset{\sim}{S}>.$$

Case 2. $Q_b < T_w < T_1$ (see Fig. 22)
Let $T_w = T_x$. As in case 1, there is a particle $\underset{\sim}{U} \in CSP<\underset{\sim}{Q},\underset{\sim}{R}>$ such that $<\underset{\sim}{U},\underset{\sim}{R},\underset{\sim}{Q},\underset{\sim}{T}>$ *after* $[Q_b]$ and

$$f_{TU} \circ f_{UT}(T_x) = T_2, \quad \text{so}$$

$$f^{-1}_{QT} \circ f^{-1}_{UQ} \circ f^{-1}_{RU} \circ f^{-1}_{QR} \circ f^{-1}_{TQ}(T_2) = T_x$$

For reasons similar to those given in case 1,

$$[T_x] \in csp<\underset{\sim}{Q},\underset{\sim}{S}>.$$

(The particle $\underset{\sim}{U}$ is not necessarily the same particle $\underset{\sim}{U}$ as in case 1. Actually there is a different particle $\underset{\sim}{U}$ for each different instant T_w).

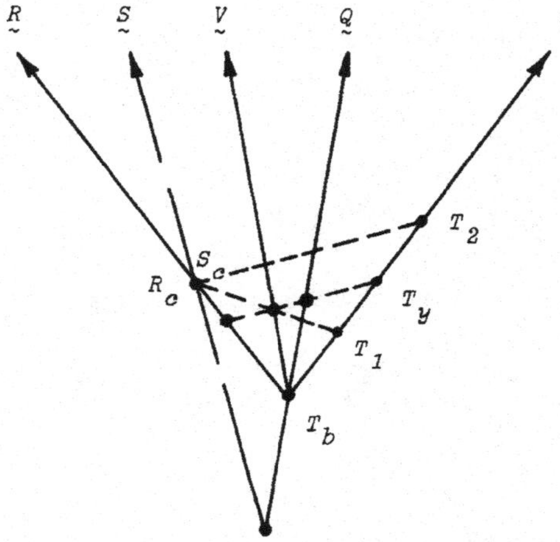

Fig. 23

Case 3. $T_1 < T_w < T_2$ (see Fig. 23)

Let $T_w = T_y$; then $T_1 < T_y < T_2$. By the previous theorem, there is a particle $\underset{\sim}{V} \varepsilon \mathit{CSP}\langle\underset{\sim}{Q},\underset{\sim}{R}\rangle$ such that $\langle\underset{\sim}{R},\underset{\sim}{V},\underset{\sim}{T}\rangle$ *after* $[Q_b]$ and

$$f_{TV} \circ f_{VT}(T_1) = T_y .$$

Since $<\underset{\sim}{R},\underset{\sim}{Q},\underset{\sim}{T}>$ *after* $[Q_b]$, and $<\underset{\sim}{R},\underset{\sim}{V},\underset{\sim}{T}>$ *after* $[Q_b]$,

$$f_{TQ} \circ f_{QR} \circ f^{-1}_{VR} \circ f^{-1}_{RV} \circ f_{RT}(T_1) = T_y \quad .$$

Therefore

$$|T_1, f^{-1}_{RV} \circ f_{RT}(T_1), R_c> ,$$

and since both

$$[T_1], [R_c] \in csp<\underset{\sim}{Q},\underset{\sim}{S}> ,$$

it follows by (1) that

$$[f^{-1}_{RV} \circ f_{RT}(T_1)] \in csp<\underset{\sim}{Q},\underset{\sim}{S}>,$$

Also, trivially

$$[f_{QR} \circ f^{-1}_{VR} \circ f^{-1}_{RV} \circ f_{RT}(T_1)] \in csp<\underset{\sim}{Q},\underset{\sim}{S}>.$$

Now either

$$<\underset{\sim}{R},\underset{\sim}{Q},\underset{\sim}{V},\underset{\sim}{T}> \textit{ after } [Q_b], \text{ or } <\underset{\sim}{R},\underset{\sim}{V},\underset{\sim}{Q},\underset{\sim}{T}> \textit{ after } [Q_b] ,$$

and in either case $[T_y]$ is in optical line with the two events

$$\left[f^{-1}_{RV} \circ f_{RT}(T_1)\right] \text{ and } \left[f_{QR} \circ f^{-1}_{VR} \circ f^{-1}_{RV} \circ f_{RT}(T_1)\right]$$

in some order, so

$$[T_y] \in csp<\underset{\sim}{Q},\underset{\sim}{S}>.$$

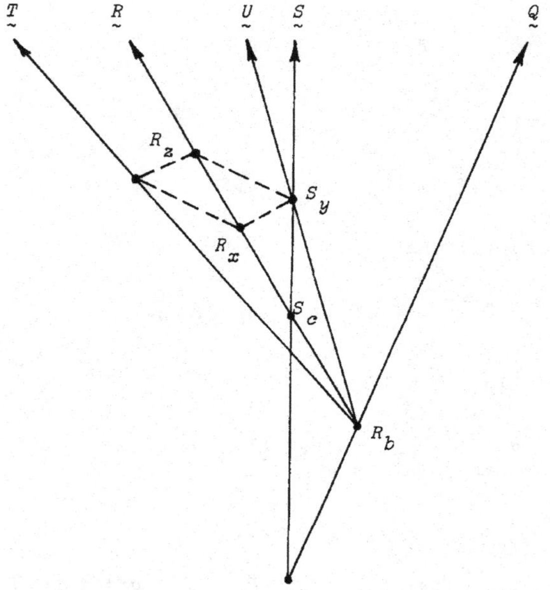

Fig. 24

Now proposition (2) applied to (3) and to cases 1, 2, 3 implies that

(4) $csp\langle \underset{\sim}{Q},\underset{\sim}{R}\rangle = csp\langle \underset{\sim}{Q},\underset{\sim}{T}\rangle \subset csp\langle \underset{\sim}{Q},\underset{\sim}{S}\rangle$.

We will now show that

(5) $$\{[S_y]: S_y > S_c, S_y \in \underset{\sim}{S}\} \subset csp\langle \underset{\sim}{Q},\underset{\sim}{R}\rangle.$$

For any instant $S_y \in \underset{\sim}{S}$ with $S_y > S_c$, let $R_x \overset{def}{=} f^{-1}_{SR}(S_y)$ and let $R_z \overset{def}{=} f_{RS}(S_y)$. At this stage we do not know whether $\langle \underset{\sim}{R},\underset{\sim}{S},\underset{\sim}{Q}\rangle$ after $[S_c]$, or whether $\underset{\sim}{R}$ and $\underset{\sim}{Q}$ are on the same side of $\underset{\sim}{S}$ *after* $[S_c]$; the former case is depicted in Fig. 24. By the previous theorem, there are particles $\underset{\sim}{T},\underset{\sim}{U} \in CSP\langle \underset{\sim}{Q},\underset{\sim}{R}\rangle$ such that $\langle \underset{\sim}{T},\underset{\sim}{R},\underset{\sim}{U}\rangle$ *after* $[R_b]$ and

$$f_{RT} \circ f_{TR}(R_x) = f_{RU} \circ f_{UR}(R_x) = f_{RS} \circ f_{SR}(R_x) = R_z.$$

By (4),

$$[f_{TR}(R_x)],\ [f_{UR}(R_x)],\ [S_y] \in csp\langle \underset{\sim}{Q},\underset{\sim}{S}\rangle,$$

so by Theorem 19 (§4.3), either

$$[f_{TR}(R_x)] = [S_y] \quad \text{or} \quad [f_{UR}(R_x)] = [S_y];$$

in either case $[S_y] \in csp\langle \underset{\sim}{Q},\underset{\sim}{R}\rangle$, which establishes (5), and hence

(6) $$csp\langle \underset{\sim}{R},\underset{\sim}{S}\rangle \subset csp\langle \underset{\sim}{Q},\underset{\sim}{R}\rangle.$$

This, together with (4), completes the proof. □

The Axiom of Incidence (Axiom IX, §2.11) is not a sufficiently strong statement for some purposes so we establish the stronger result contained in the following:

THEOREM 28 (Ordered Incidence - see Fig. 25)

Let $\underset{\sim}{R}, \underset{\sim}{S}$ *be distinct particles with instants* $R_c \in \underset{\sim}{R}$, $S_c \in \underset{\sim}{S}$ *such that* $R_c \simeq S_c$.
Given an instant $S_a \in \underset{\sim}{S}$ *with* $S_a < S_c$, *there is a particle* $\underset{\sim}{Q}$ *with instants* $Q_a, Q_b \in \underset{\sim}{Q}$ *and there is an instant* $R_b \in \underset{\sim}{R}$ *such that*

$$S_a \simeq Q_a < Q_b \simeq R_b < R_c \simeq S_c \ .$$

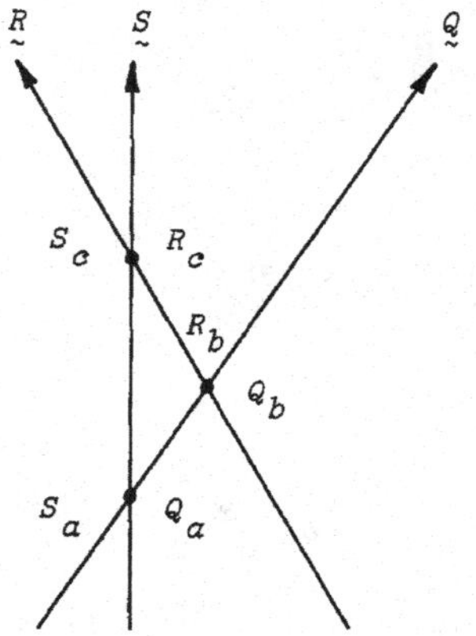

Fig. 25

This theorem is a consequence of Axioms IV (§2.4), VI (§2.8), IX (§2.11) and Theorems 17 (§4.2), 19 (§4.3), 26 (§5.5) and 27 (§6.1). It is used in the proof of Theorem 30 (§6.3), Corollary 1 to Theorem 30 (§6.3), and Theorems 32 (§6.4) and 33 (§6.4).

PROOF. By the Axiom of Incidence (Axiom IX, §2.11), there is a particle $\underset{\sim}{W}$ with instants $W_a, W_d \in \underset{\sim}{W}$ and there is an instant $R_d \in \underset{\sim}{R}$ such that

$$W_a \simeq S_a \text{ and } W_d \simeq R_d \neq R_c.$$

Case 1. $W_a < W_d < S_c$ (see Fig. 25)
In this trivial case we let $\underset{\sim}{Q} \overset{def}{=} \underset{\sim}{W}$, $Q_a \overset{def}{=} W_a$, $Q_b \overset{def}{=} W_d$ and $R_b \overset{def}{=} R_d$.
Case 2. $W_d < W_a$ (see Fig. 26)
Consider an instant $S_b \in \underset{\sim}{S}$ with $S_a < S_b < S_c$. By Theorem 26 (§5.5), there are particles $\underset{\sim}{T}, \underset{\sim}{U} \in CSP\langle\underset{\sim}{W}, \underset{\sim}{S}\rangle$ such that $\langle\underset{\sim}{W}, \underset{\sim}{S}, \underset{\sim}{T}\rangle$ *after* $[S_a]$ and $\langle\underset{\sim}{U}, \underset{\sim}{S}, \underset{\sim}{T}\rangle$ *after* $[S_a]$ and

(1) $$f_{ST} \circ f_{TS}(S_b) = f_{SR} \circ f_{RS}(S_b) = f_{SU} \circ f_{US}(S_b).$$

By the previous theorem,

$$[f_{TS}(S_b)], [f_{US}(S_b)] \in csp\langle\underset{\sim}{W}, \underset{\sim}{S}\rangle \subset csp\langle\underset{\sim}{W}, \underset{\sim}{R}\rangle.$$

By equation (1) and Theorem 19 (§4.3), either $\underset{\sim}{T}$ or $\underset{\sim}{U}$ coincides with $\underset{\sim}{R}$ at $[f_{RS}(S_b)]$; so, accordingly, we define $\underset{\sim}{Q}$ to be either $\underset{\sim}{T}$ or $\underset{\sim}{U}$ so that $\underset{\sim}{Q}$ coincides with $\underset{\sim}{R}$ at $[f_{RS}(S_b)]$. We then define

$R_b \overset{def}{=} f_{RS}(S_b)$ and $Q_b \overset{def}{=} f_{QR}(R_b)$, and so $Q_b \simeq R_b$.

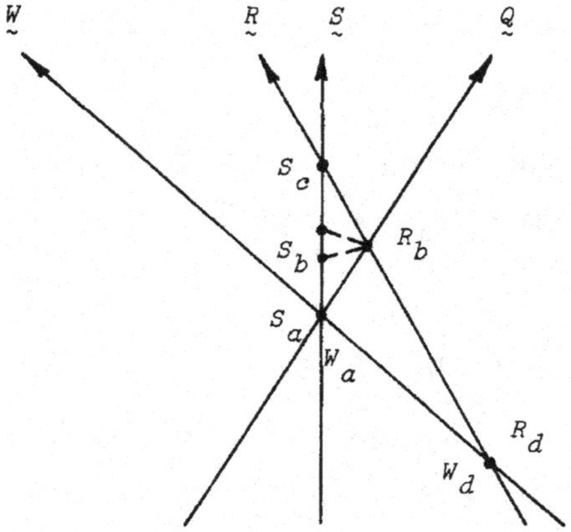

Fig. 26

Case 3. $S_c < W_d$ (see Fig. 27)

We first find a particle $\underset{\sim}{T}$ which has properties as stated in the paragraph containing (4).

Case 3(i). If $f_{WS} \circ f_{SR} \circ f_{RS}(S_a) \leq f_{SW}^{-1}(S_c)$, we let $\underset{\sim}{W}$ be the particle $\underset{\sim}{T}$; that is we define $\underset{\sim}{T} \overset{def}{=} \underset{\sim}{W}$, $T_a \overset{def}{=} W_a$, $T_3 \overset{def}{=} W_d$, and $R_3 \overset{def}{=} R_d$.

Case 3(ii). If $f_{WS} \circ f_{SR} \circ f_{RS}(S_a) > f_{SW}^{-1}(S_c)$, by Theorem 26 (§5.5) there is a particle $\underset{\sim}{T} \in CSP\langle\underset{\sim}{W},\underset{\sim}{S}\rangle$ with either $\langle\underset{\sim}{W},\underset{\sim}{T},\underset{\sim}{S}\rangle$ *after* $[S_a]$, or $\langle\underset{\sim}{W},\underset{\sim}{S},\underset{\sim}{T}\rangle$ *after* $[S_a]$, such that

(2) $\quad f_{WT} \circ f_{TW} \circ f^{-1}_{SW}(S_c) = f_{WS} \circ f_{SR} \circ f_{RS}(S_a) > f^{-1}_{SW}(S_c)$.

Since $\underset{\sim}{R}$ coincides with $\underset{\sim}{S}$ at $[S_c] > [S_a]$,

$$f_{SR} \circ f_{RS}(S_a) < S_c ,$$

whence by (2),

$$f_{WT} \circ f_{TW} \circ f^{-1}_{SW}(S_c) < f_{WS} \circ f_{SW} \circ f^{-1}_{SW}(S_c)$$

and therefore $<\underset{\sim}{W},\underset{\sim}{T},\underset{\sim}{S}>$ *after* $[S_a]$ (see Fig. 27). Hence, by (2),

$$f_{WT} \circ f^{-1}_{ST}(S_c) = f_{WT} \circ f_{TS} \circ f_{SR} \circ f_{RS}(S_a)$$

and so

(3) $\quad T_1 \overset{def}{=} f^{-1}_{ST}(S_c) = f_{TS} \circ f_{SR} \circ f_{RS}(S_a)$.

Now $<\underset{\sim}{W},\underset{\sim}{T},\underset{\sim}{S}>$ *after* $[S_a]$; $\underset{\sim}{R}$ coincides with $\underset{\sim}{S}$ at $[S_c]$ and with $\underset{\sim}{W}$ at $[W_d]$; so by Theorem 17 (§4.2) there is an instant $T_3 \in \underset{\sim}{T}$ such that $\underset{\sim}{R}$ coincides with $\underset{\sim}{T}$ at the event $[T_3]$

$$S_c < T_3 < W_d .$$

Let

$$R_3 \overset{def}{=} f_{RT}(T_3) \simeq T_3$$

and let

$$T_a \overset{def}{=} f_{TS}(S_a) \simeq S_a .$$

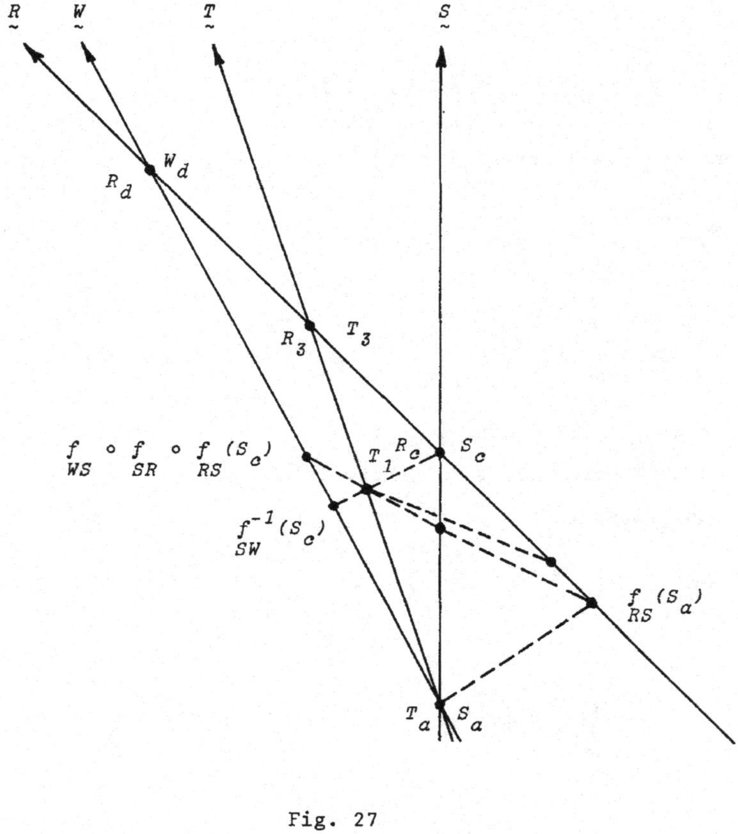
R
W
T
S
W_d
R_d
R_3
T_3
$f_{WS} \circ f_{SR} \circ f_{RS}(S_c)$
T_1
R_c
S_c
$f_{SW}^{-1}(S_c)$
$f_{RS}(S_a)$
T_a
S_a

Fig. 27

So for both cases 3(i) and 3(ii): *there is a particle* $\underset{\sim}{T}$ *and instants* $T_a, T_3 \in \underset{\sim}{T}$ *and* $R_3 \in \underset{\sim}{R}$ *such that* (see Fig. 28),

$$T_a \simeq S_a < S_c \simeq R_c < R_3 \simeq T_3 \quad \textit{and}$$

$$f_{TS} \circ f_{SR} \circ f_{RS}(S_a) \leq f_{ST}^{-1}(S_c) \overset{def}{=} T_1 .$$

Then, by the Triangle Inequality (Axiom IV, §2.4),

(4) $$f_{RS}(S_a) \leq f_{SR}^{-1} \circ f_{TS}^{-1} \circ f_{ST}^{-1}(S_c) \leq f_{TR}^{-1}(T_1) .$$

By the Axiom of the Intermediate Particle (Axiom VI, §2.8), there is a particle $\underset{\sim}{U} \in CSP\langle\underset{\sim}{R},\underset{\sim}{T}\rangle$ with $\langle\underset{\sim}{T},\underset{\sim}{U},\underset{\sim}{R}\rangle$ and $\underset{\sim}{T} \neq \underset{\sim}{U} \neq \underset{\sim}{R}$. Let $U_2 \overset{def}{=} f_{RU}^{-1}(R_c)$ and $T_2 \overset{def}{=} f_{TU}(U_2)$ and let $R_2 \overset{def}{=} f_{UR}^{-1}(U_2)$ (see Fig. 28). Then

(5) $$|T_1, U_2, R_c\rangle \quad \text{and} \quad |R_2, U_2, T_2\rangle \; .$$

By Theorem 26 (§5.5), there is a particle $\underset{\sim}{V} \in CSP\langle\underset{\sim}{T},\underset{\sim}{S}\rangle$ such that $\langle\underset{\sim}{T},\underset{\sim}{V},\underset{\sim}{S}\rangle$ *after* $[S_a]$ and

(6) $$f_{TV} \circ f_{VT}(T_1) = f_{TU} \circ f_{UT}(T_1) = T_2 .$$

Letting $V_2 \overset{def}{=} f_{SV}^{-1}(S_c)$ we have $|T_1, V_2, S_c\rangle$ and, from (5), $|T_1, U_2, S_c\rangle$. Thus by Theorem 19 (§4.3) and (6), $U_2 \simeq V_2$, so by (5), $|R_2, V_2, T_2\rangle$; that is

$$[R_2] \in csp\langle\underset{\sim}{T},\underset{\sim}{V}\rangle = csp\langle\underset{\sim}{T},\underset{\sim}{S}\rangle .$$

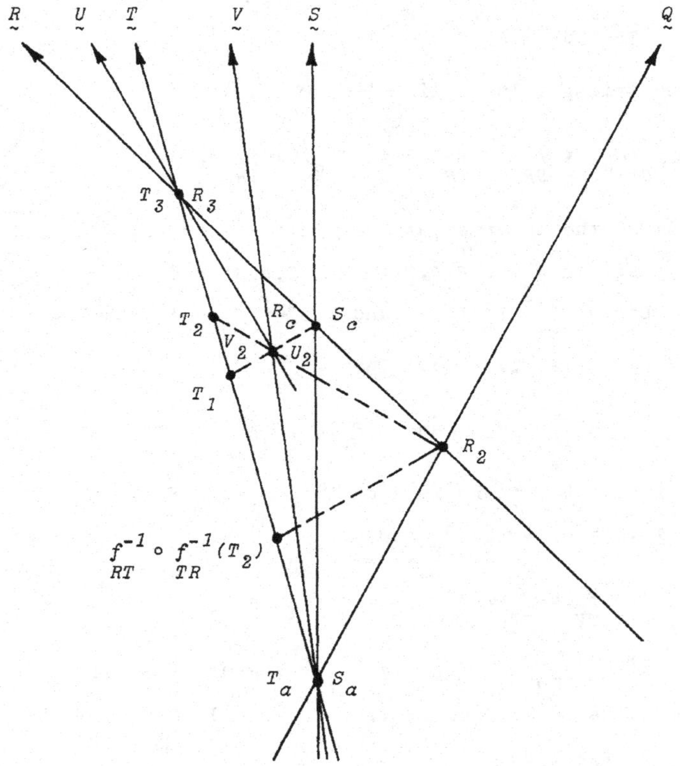

Fig. 28

Since $T_2 > T_1$, it follows from (4) that

$$S_a \simeq T_a \leqslant f^{-1}_{RT} \circ f^{-1}_{TR}(T_1) < f^{-1}_{RT} \circ f^{-1}_{TR}(T_2).$$

By Theorem 26 (§5.5), there is a particle $\underset{\sim}{Q} \in CSP\langle \underset{\sim}{T}, \underset{\sim}{S}\rangle$ such that $\langle \underset{\sim}{T}, \underset{\sim}{S}, \underset{\sim}{Q}\rangle$ *after* $[T_a]$ and

$$f^{-1}_{QT} \circ f^{-1}_{TQ}(T_2) = f^{-1}_{RT} \circ f^{-1}_{TR}(T_2) \;.$$

By Theorem 19 (§4.3), $\underset{\sim}{Q}$ coincides with $\underset{\sim}{R}$ at $[R_2]$. We now let $R_b \overset{def}{=} R_2$, $Q_b \overset{def}{=} f_{QR}(R_b)$; so that $Q_b \simeq R_b$. □

§6.2 The Crossing Theorem

Given two distinct particles $\underset{\sim}{Q}, \underset{\sim}{T}$ with instants $Q_c \in \underset{\sim}{Q}$ and $T_c \in \underset{\sim}{T}$ such that $Q_c \simeq T_c$, we partition the events of $csp\langle \underset{\sim}{Q}, \underset{\sim}{T}\rangle$ into two classes, called *left and right sides*, by specifying that events after $[Q_c]$ and coincident with $\underset{\sim}{T}$ are on one particular side of $\underset{\sim}{Q}$. Since distinct particles coincide at no more than one event (Theorem 6, §2.9), this definition is in accordance with the conclusions of §4.1 and §4.2. Instead of saying that "events after $[Q_c]$ and coincident with $\underset{\sim}{T}$ are on the left (right) side of $\underset{\sim}{Q}$ in $csp\langle \underset{\sim}{Q}, \underset{\sim}{T}\rangle$", we shall simply say that *"$\underset{\sim}{T}$ is on the left (right) side of $\underset{\sim}{Q}$ after $[Q_c]$ in $csp\langle \underset{\sim}{Q}, \underset{\sim}{T}\rangle$*. Sometimes if the context is unambiguous, we shall delete the words "in $csp\langle \underset{\sim}{Q}, \underset{\sim}{T}\rangle$". At this stage we do not define the sides of $\underset{\sim}{Q}$ before $[Q_c]$; they are specified completely by the corollary to Theorem 33 (§6.4).

Let $\underset{\sim}{U}$ be a particle with an instant $U_d \in \underset{\sim}{U}$, with $U_d > Q_c$, such that $\underset{\sim}{U}$ coincides with $\underset{\sim}{Q}$ at $[U_d]$. If

(i) $[\underset{\sim}{Q},\underset{\sim}{T},\underset{\sim}{U}]$ *after* $[Q_c]$,

(ii) $\underset{\sim}{U}$ is on the left (right) side of $\underset{\sim}{Q}$ before $[U_d]$ in $csp\langle\underset{\sim}{Q},\underset{\sim}{T}\rangle$ and

(iii) $\underset{\sim}{U}$ is on the right (left) side of $\underset{\sim}{Q}$ after $[U_d]$ in $csp\langle\underset{\sim}{Q},\underset{\sim}{T}\rangle$,

we say that *$\underset{\sim}{U}$ crosses $\underset{\sim}{Q}$ at $[U_d]$ in $csp\langle\underset{\sim}{Q},\underset{\sim}{T}\rangle$*, or simply that *$\underset{\sim}{U}$ crosses $\underset{\sim}{Q}$ at $[U_d]$* if the context of the statement is unambiguous (see Fig. 29).

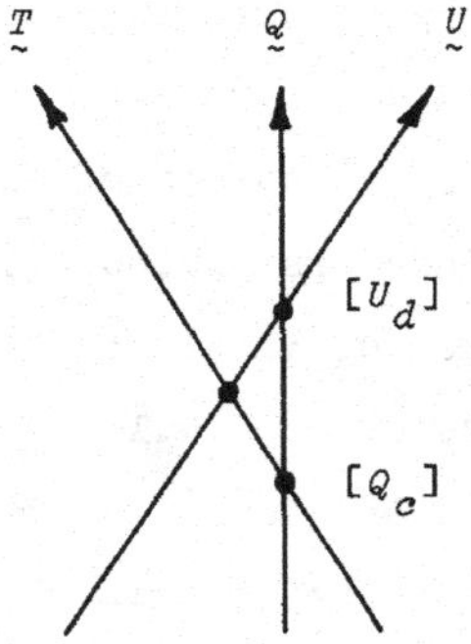

Fig. 29

THEOREM 29 (Crossing Theorem)

Let $\underset{\sim}{Q}, \underset{\sim}{R}, \underset{\sim}{S}$ *be distinct particles with instants* $Q_a, Q_b \in \underset{\sim}{Q}$; $R_b, R_c \in \underset{\sim}{R}$; $S_a, S_c \in \underset{\sim}{S}$ *such that*

$$S_a \simeq Q_a < Q_b \simeq R_b < R_c \simeq S_c.$$

Then $\underset{\sim}{R}$ *crosses* $\underset{\sim}{S}$ *at* $[S_c]$ *in* $csp\langle\underset{\sim}{Q},\underset{\sim}{R}\rangle$.

This theorem is a consequence of Theorems 6 (§2.9), 17 (§4.2), 19 (§4.3), 26 (§5.5) and 27 (§6.1). It is used in the proofs of Theorem 30 (§6.3), Corollary 1 to Theorem 30 (§6.3) and Theorems 32 (§6.4) and 33 (§6.4).

PROOF. (see Fig. 30)

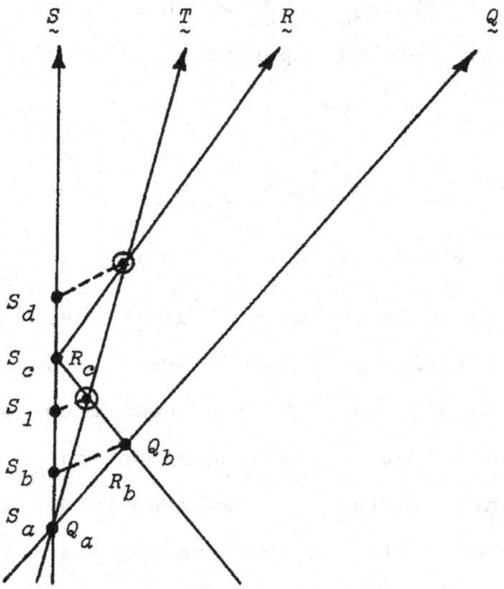

Fig. 30

Let the right side of $\underset{\sim}{S}$ (in $csp\langle\underset{\sim}{Q},\underset{\sim}{S}\rangle$ be defined such that $\underset{\sim}{Q}$ is on the right side of $\underset{\sim}{S}$ *after* $[S_a]$.
By Theorem 27 (6.1),

$$[\underset{\sim}{Q},\underset{\sim}{R},\underset{\sim}{S}] \quad after \quad [R_b].$$

Let us suppose that $\underset{\sim}{R}$ does not cross $\underset{\sim}{S}$ at $[S_c]$ and we shall deduce a contradiction. If $\underset{\sim}{R}$ does not cross $\underset{\sim}{S}$ at $[S_c]$, it follows that $\underset{\sim}{R} \setminus \{R_c\}$ is on the right side of $\underset{\sim}{S}$ after $[R_b]$. Consider any instant $S_d \in \underset{\sim}{S}$ such that $S_d > S_c$ and

$$f_{SR} \circ f_{RS}(S_d) < f_{SQ} \circ f_{QS}(S_d).$$

(This is possible since otherwise $\underset{\sim}{Q}$ would coincide with $\underset{\sim}{S}$ at $[S_c]$ and so by Theorem 6 (§2.9), $\underset{\sim}{Q} \approx \underset{\sim}{S}$, which is a contradiction). By Theorem 26 (§5.5), there is a particle $\underset{\sim}{T} \in CSP\langle\underset{\sim}{Q},\underset{\sim}{S}\rangle$ such that $\langle\underset{\sim}{S},\underset{\sim}{T},\underset{\sim}{Q}\rangle$ *after* $[Q_a]$ and

$$f_{ST} \circ f_{TS}(S_d) = f_{SR} \circ f_{RS}(S_d),$$

whence by Theorem 19 (§4.3), $\underset{\sim}{T}$ coincides with $\underset{\sim}{R}$ at $[f_{RS}(S_d)]$.
The particle $\underset{\sim}{R}$ is on the left side of $\underset{\sim}{T}$ at the event $[R_c]$ and it is on the right side of $\underset{\sim}{T}$ at the event $[R_b]$ so by Theorem 17 (§4.2), $\underset{\sim}{R}$ coincides with $\underset{\sim}{T}$ at some event before $[R_c]$ (and after $[R_b]$). But now we have shown that $\underset{\sim}{R}$ coincides with $\underset{\sim}{T}$ at two distinct events, which is a contradiction by Theorem 6 (§2.9). We conclude that $\underset{\sim}{R}$ crosses $\underset{\sim}{S}$ at $[S_c]$. □

§6.3 Collinearity of Three Particles. Properties of Collinear Sub-SPRAYs.

In absolute geometry any three lines are coplanar if they intersect in any three distinct points. We now prove the analogous:

THEOREM 30 (Second Collinearity Theorem)

Any three particles which coincide (in pairs) at three distinct events are (permanently) collinear.

This theorem is a consequence of Theorems 5 (§2.9), 19 (§4.3), 26 (§5.5), 27 (§6.1), 28 (§6.1) and 29 (§6.2). It is used in the proofs of Corollaries 1 and 2 of Theorem 32 (§6.4).

PROOF. (see Fig. 31)

Let $\underset{\sim}{Q}, \underset{\sim}{R}, \underset{\sim}{S}$ be distinct particles with instants $Q_a, Q_b \in \underset{\sim}{Q}$; $R_b, R_c \in \underset{\sim}{R}$; $S_a, S_c \in \underset{\sim}{S}$ such that

$$S_a \simeq Q_a < Q_b \simeq R_b < R_c \simeq S_c .$$

Take any instant $S_\gamma \in \underset{\sim}{S}$ with $S_\gamma < S_a$. We will show that

$$[\underset{\sim}{Q}, \underset{\sim}{R}, \underset{\sim}{S}] \text{ after } [S_\gamma] .$$

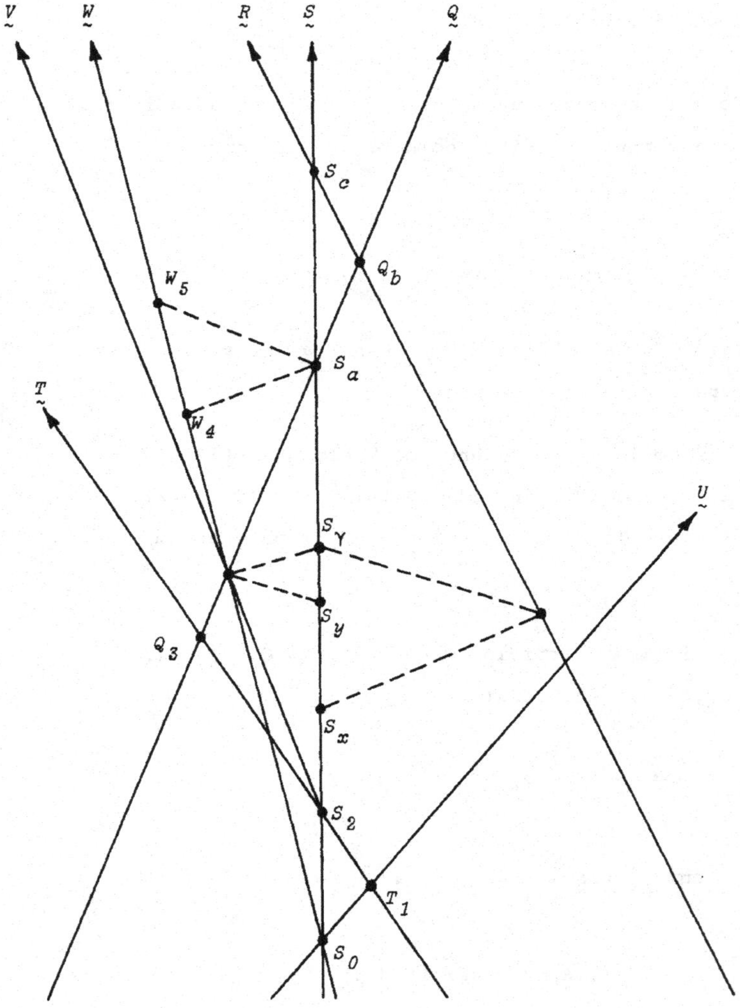

Fig. 31

Let $S_x \overset{def}{=} f^{-1}_{RS} \circ f^{-1}_{SR}(S_\Upsilon)$, let $S_y \overset{def}{=} f^{-1}_{QS} \circ f^{-1}_{SQ}(S_\Upsilon)$ and take any two instants $S_0, S_2 \in \underset{\sim}{S}$ such that

$$S_0 < S_2 < min\{S_x, S_y\}.$$

By Theorem 28 (§6.1), there is a particle $\underset{\sim}{T}$ with instants $T_2, T_3 \in \underset{\sim}{T}$ and there is an instant $Q_3 \in \underset{\sim}{Q}$ such that

$$S_2 \simeq T_2 < T_3 \simeq Q_3 < Q_a \simeq S_a.$$

Again by Theorem 28 (§6.1), there is a particle $\underset{\sim}{U}$ with instants $U_0, U_1 \in \underset{\sim}{U}$ and there is an instant $T_1 \in \underset{\sim}{T}$ such that

$$S_0 \simeq U_0 < U_1 \simeq T_1 < T_2 \simeq S_2.$$

By Theorem 27 (§6.1),

(1)
$$\begin{aligned} csp\langle \underset{\sim}{S}, \underset{\sim}{U}\rangle &\supset csp\langle \underset{\sim}{T}, \underset{\sim}{U}\rangle \\ &\supset csp\langle \underset{\sim}{S}, \underset{\sim}{T}\rangle \\ &\supset csp\langle \underset{\sim}{Q}, \underset{\sim}{T}\rangle \\ &\supset csp\langle \underset{\sim}{Q}, \underset{\sim}{S}\rangle \\ &\supset csp\langle \underset{\sim}{Q}, \underset{\sim}{R}\rangle \\ &\supset csp\langle \underset{\sim}{R}, \underset{\sim}{S}\rangle. \end{aligned}$$

We define the right side of $\underset{\sim}{S}$ in $csp\langle \underset{\sim}{S}, \underset{\sim}{U}\rangle$ to be the side which contains $\underset{\sim}{U}$ after $[S_0]$. By Theorem 26 (§5.5) there are particles

(2) $\underset{\sim}{V} \in CSP\langle \underset{\sim}{S}, \underset{\sim}{T}\rangle$ and $\underset{\sim}{W} \in CSP\langle \underset{\sim}{S}, \underset{\sim}{U}\rangle$
with $\underset{\sim}{V}$ on the left side of $\underset{\sim}{S}$ after $[S_2]$ and
with $\underset{\sim}{W}$ on the left side of $\underset{\sim}{S}$ after $[S_0]$ such that

(3) $$f_{SV} \circ f_{VS}(S_y) = f_{SW} \circ f_{WS}(S_y) = S_Y.$$

Thus by Theorem 19 (§4.3), $\underset{\sim}{V}$ and $\underset{\sim}{W}$ coincide at the event $[f_{VS}(S_y)]$. By Theorem 26 (§5.5) and Theorem 27 (§6.1),

(4) $$csp\langle \underset{\sim}{S}, \underset{\sim}{U}\rangle \supset csp\langle \underset{\sim}{V}, \underset{\sim}{W}\rangle.$$

Let $W_4 \stackrel{def}{=} f^{-1}_{SW}(S_a)$, let $W_5 \stackrel{def}{=} f_{WS}(S_a)$; then by Theorem 26 (§5.5), there is a particle $\underset{\sim}{Q}^* \in CSP\langle \underset{\sim}{V}, \underset{\sim}{W}\rangle$ with $\underset{\sim}{Q}^*$ on the right side of $\underset{\sim}{W}$ after $[f_{VS}(S_y)]$ such that

(5) $$f_{WQ^*} \circ f_{Q^*W}(W_4) = W_5.$$

By repeated application of the previous theorem to the crossing of the pairs $(\underset{\sim}{T}, \underset{\sim}{U})$, $(\underset{\sim}{T}, \underset{\sim}{S})$, $(\underset{\sim}{T}, \underset{\sim}{Q})$, $(\underset{\sim}{S}, \underset{\sim}{Q})$, $(\underset{\sim}{R}, \underset{\sim}{Q})$, $(\underset{\sim}{R}, \underset{\sim}{S})$ and $(\underset{\sim}{V}, \underset{\sim}{W})$, $(\underset{\sim}{W}, \underset{\sim}{Q}^*)$, we see that after coincidence (of each pair), the first particle (of each pair) is on the left side of the second particle. So by (5) and Theorem 19 (§4.3), the particles $\underset{\sim}{Q}$ and $\underset{\sim}{Q}^*$ coincide at $[S_a]$; that is

$$\underset{\sim}{Q}, \underset{\sim}{Q}^* \in SPR[S_a]$$

and $\underset{\sim}{Q}^*$ is on the right side of $\underset{\sim}{S}$ *after* $[S_a]$. By (3), the definitions of $\underset{\sim}{Q}^*$ and S_y, and Theorem 5 (§2.9),

$$f_{SQ} \circ f_{QS} = f_{SQ^*} \circ f_{Q^*S}.$$

Since both $\underset{\sim}{Q}$ and $\underset{\sim}{Q}^*$ are on the right side of $\underset{\sim}{S}$ after $[S_a]$, Theorem 19 (§4.3) implies that $\underset{\sim}{Q}^* \simeq \underset{\sim}{Q}$. Hence

(6) $$\{[Q_z]\colon Q_z > S_y,\ Q_z \in \underset{\sim}{Q}\} \subset csp\langle \underset{\sim}{S}, \underset{\sim}{U}\rangle.$$

By a similar procedure (based on the third and subsequent paragraphs), we can show that

(7) $$\{[R_z]\colon R_z > S_x,\ R_z \in \underset{\sim}{R}\} \subset csp\langle \underset{\sim}{S}, \underset{\sim}{U}\rangle.$$

Thus from (6) and (7)

$$[\underset{\sim}{Q}, \underset{\sim}{R}, \underset{\sim}{S}]\ \mathit{after}\ [S_\gamma]\ .$$

But S_γ was arbitrary; hence $[\underset{\sim}{Q}, \underset{\sim}{R}, \underset{\sim}{S}]$, which completes the proof. □

COROLLARY 1. *Let* $\underset{\sim}{Q}, \underset{\sim}{R}, \underset{\sim}{S}$ *be distinct particles which coincide at the event* $[Q_c]$.
If $\langle \underset{\sim}{Q}, \underset{\sim}{R}, \underset{\sim}{S}\rangle$ *after* $[Q_c]$, *then* $\langle \underset{\sim}{Q}, \underset{\sim}{R}, \underset{\sim}{S}\rangle$.

This corollary is a consequence of Theorems 4 (§2.7), 6 (§2.9), 17 (§4.2), 27 (§6.1), 28 (§6.1), and 29 (§6.2). It is used in the proof of Theorem 31 (§6.3).

PROOF. The previous method of proof applies here, but with the first set of order and coincidence relations being replaced by

$$S_c \simeq Q_c = Q_c \simeq R_c = R_c \simeq S_c \quad \text{and}$$

$$\langle \underset{\sim}{Q}, \underset{\sim}{R}, \underset{\sim}{S}\rangle\ \mathit{after}\ [Q_c];$$

the second statement is required to justify the last three of relations (1) of the previous proof. No further changes are required to show that $[\underset{\sim}{Q},\underset{\sim}{R},\underset{\sim}{S}]$.

By Theorem 17 (§4.2) and Theorem 6 (§2.9) either (i) $\langle\underset{\sim}{Q},\underset{\sim}{S},\underset{\sim}{R}\rangle$ *before* $[Q_c]$, or (ii) $\langle\underset{\sim}{S},\underset{\sim}{Q},\underset{\sim}{R}\rangle$ *before* $[Q_c]$, or (iii) $\langle\underset{\sim}{Q},\underset{\sim}{R},\underset{\sim}{S}\rangle$ *before* $[Q_c]$. We will show that case (iii) is true by showing that each of cases (i) and (ii) lead to contradictions.

Case (i) $\langle\underset{\sim}{Q},\underset{\sim}{S},\underset{\sim}{R}\rangle$ *before* $[Q_c]$ (see Fig. 32)
Take any instant $Q_1 \in \underset{\sim}{Q}$ with $Q_1 < Q_c$. By Theorem 28 (§6.1) there is a particle $\underset{\sim}{T}$ and instants $T_1, T_3 \in \underset{\sim}{T}$ and $R_3 \in \underset{\sim}{R}$ such that (see Fig. 32)

$$Q_1 \approx T_1 < T_3 \approx R_3 < R_c \approx Q_c.$$

It follows from the above theorem that $[\underset{\sim}{Q},\underset{\sim}{R},\underset{\sim}{T}]$ and since $\langle\underset{\sim}{Q},\underset{\sim}{R},\underset{\sim}{S}\rangle$ it follows from Theorem 4 (§2.7) that $[\underset{\sim}{Q},\underset{\sim}{R},\underset{\sim}{S},\underset{\sim}{T}]$. By Theorem 17 (§4.2) there is an instant $T_2 \in \underset{\sim}{T}$ with $T_1 < T_2 < T_3$ such that $\underset{\sim}{T}$ coincides with $\underset{\sim}{S}$ at $[T_2]$. By Theorem 27 (§6.1)

$\mathit{csp}\langle\underset{\sim}{Q},\underset{\sim}{T}\rangle \supset \mathit{csp}\langle\underset{\sim}{S},\underset{\sim}{T}\rangle \supset \mathit{csp}\langle\underset{\sim}{Q},\underset{\sim}{S}\rangle$,
$\mathit{csp}\langle\underset{\sim}{Q},\underset{\sim}{T}\rangle \supset \mathit{csp}\langle\underset{\sim}{R},\underset{\sim}{T}\rangle \supset \mathit{csp}\langle\underset{\sim}{Q},\underset{\sim}{R}\rangle$, and
$\mathit{csp}\langle\underset{\sim}{S},\underset{\sim}{T}\rangle \supset \mathit{csp}\langle\underset{\sim}{R},\underset{\sim}{T}\rangle \supset \mathit{csp}\langle\underset{\sim}{R},\underset{\sim}{S}\rangle$.
Thus by Theorem 29 (§6.2),
$\underset{\sim}{S}$ crosses $\underset{\sim}{Q}$ at $[Q_c]$ in $\mathit{csp}\langle\underset{\sim}{Q},\underset{\sim}{T}\rangle$
$\underset{\sim}{R}$ crosses $\underset{\sim}{Q}$ at $[Q_c]$ in $\mathit{csp}\langle\underset{\sim}{Q},\underset{\sim}{T}\rangle$, and
$\underset{\sim}{R}$ crosses $\underset{\sim}{S}$ at $[Q_c]$ in $\mathit{csp}\langle\underset{\sim}{Q},\underset{\sim}{T}\rangle$; whence $\langle\underset{\sim}{Q},\underset{\sim}{R},\underset{\sim}{S}\rangle$ *before* $[Q_c]$ which is a contradiction.

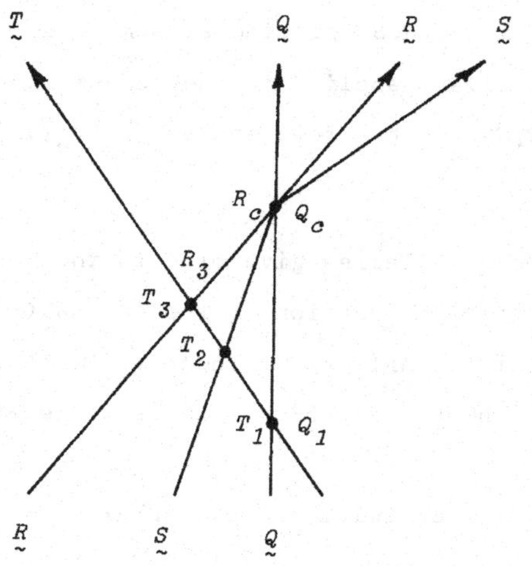

Fig. 32

Case (ii) $<\underset{\sim}{S},\underset{\sim}{Q},\underset{\sim}{R}>$ *before* $[Q_c]$.
Similarly, by interchanging the symbols Q and S wherever they occur in the above paragraph, we deduce a contradiction. (Note that Fig. 32 does not apply to this case, even with Q and S interchanged).

The only remaining possibility is case (iii); that is $<\underset{\sim}{Q},\underset{\sim}{R},\underset{\sim}{S}>$ *before* $[Q_c]$. □

COROLLARY 2. *If* $\underset{\sim}{S} \in \hat{Q}_{\underset{\sim}{R}}$ *then* $<\underset{\sim}{Q},\underset{\sim}{R},\underset{\sim}{S}>$.
(This is a stronger statement than the definition of §5.3).

This corollary is a consequence of the previous corollary. It is used in the proofs of Theorem 31 (§6.3) and Corollary 2 of Theorem 32 (§6.4).

PROOF. If the particles $\underset{\sim}{Q},\underset{\sim}{R},\underset{\sim}{S}$ coincide at some event $[Q_c]$, then the previous corollary applied to the definition of §5.3 shows that $\langle \underset{\sim}{Q},\underset{\sim}{R},\underset{\sim}{S} \rangle$ (both before and after $[Q_c]$). □

The previous two corollaries give rise to the following theorem, which is a stronger version of Theorem 26 (§5.5); a more general statement of part (v) of Theorem 26 (§5.5) appears as a special case of Theorem 33 (§6.4) and is not stated here.

THEOREM 31 (Properties of Collinear sub-SPRAYs)

Let $\underset{\sim}{Q},\underset{\sim}{R}$ *and* $\underset{\sim}{S},\underset{\sim}{T}$ *be pairs of distinct particles coinciding at the event* $[Q_c]$. *Then*

(i) $CSP\langle \underset{\sim}{Q},\underset{\sim}{R} \rangle$ *is a collinear set of particles, and*

$$CSP\langle \underset{\sim}{Q},\underset{\sim}{R} \rangle = \{S: \ \langle \underset{\sim}{S},\underset{\sim}{Q},\underset{\sim}{R} \rangle, \ \langle \underset{\sim}{Q},\underset{\sim}{S},\underset{\sim}{R} \rangle, \text{ or } \langle \underset{\sim}{Q},\underset{\sim}{R},\underset{\sim}{S} \rangle \ ; \ \underset{\sim}{S} \in SPR[Q_c]\} ,$$

(ii) $\underset{\sim}{S},\underset{\sim}{T} \in CSP\langle \underset{\sim}{Q},\underset{\sim}{R} \rangle \Longrightarrow CSP\langle \underset{\sim}{S},\underset{\sim}{T} \rangle = CSP\langle \underset{\sim}{Q},\underset{\sim}{R} \rangle$,

(iii) $\underset{\sim}{S},\underset{\sim}{T} \in CSP\langle \underset{\sim}{Q},\underset{\sim}{R} \rangle \Longrightarrow \hat{\underset{\sim}{S}}_T, \hat{\underset{\sim}{T}}_S \subset CSP\langle \underset{\sim}{Q},\underset{\sim}{R} \rangle$, and

(iv) $CSP\langle \underset{\sim}{Q},\underset{\sim}{R} \rangle$ *is a simply ordered set.*

This theorem is a consequence of Theorem 26 (§5.5) and Corollaries 1 and 2 of Theorem 30 (§6.3). It is used in the proof of Theorem 32 (§6.4).

PROOF. Part (i) is a consequence of the previous two corollaries applied to Theorem 26 (§5.5).
Part (ii) and part (iii) are the same propositions as (ii) and (iii) of Theorem 26 (§5.5).
In the case of part (iv) we define an order relation "to the right of" after [Q_c], which is specified by defining the particle $\underset{\sim}{S}$ to be on the right side of $\underset{\sim}{Q}$ after [Q_c] in $csp\langle \underset{\sim}{Q}, \underset{\sim}{S}\rangle$. Before the event of coincidence [Q_c], the particles have the same ordering, but the ordering does not correspond to a relation of "to the right of", but rather to a relation of "to the left of" (see the corollary to Theorem 33 (§6.4)). □

§6.4 Properties of Collinear Sets of Particles

We now prove theorems which are analogues of the following propositions which occur in absolute geometry:

(i) given any two distinct points in a given plane, there is a line in the plane containing both points (Theorem 33, §6.4),

(ii) a line and a point not on the line determine a plane (Corollary 2 of Theorem 33, §6.4) and

(iii) any line which intersects a plane in two distinct points is contained in the plane (Theorem 34, §6.4).

For any pair of distinct particles $\underset{\sim}{Q},\underset{\sim}{S}$ which coincide at some event $[Q_c]$, we define

$$COL[\underset{\sim}{Q},\underset{\sim}{S}] \overset{def}{=} \{\underset{\sim}{R}: [\underset{\sim}{R},\underset{\sim}{Q},\underset{\sim}{S}],\ \underset{\sim}{R} \in \mathcal{P}\};$$

that is, $COL[\underset{\sim}{Q},\underset{\sim}{S}]$ is the *set of particles which are collinear with* $\underset{\sim}{Q}$ *and* $\underset{\sim}{S}$. We also define

$$col[\underset{\sim}{Q},\underset{\sim}{S}] \overset{def}{=} \{[R_x]: R_x \in \underset{\sim}{R}, \underset{\sim}{R} \in COL[\underset{\sim}{Q},\underset{\sim}{S}]\}.$$

We note as a result of the previous theorem that:

$$CSP\langle\underset{\sim}{Q},\underset{\sim}{S}\rangle = COL[\underset{\sim}{Q},\underset{\sim}{S}] \cap SPR[Q_c] \subset COL[\underset{\sim}{Q},\underset{\sim}{S}] \text{ and}$$
$$csp\langle\underset{\sim}{Q},\underset{\sim}{S}\rangle = col[\underset{\sim}{Q},\underset{\sim}{S}] \cap \{[R_x]: R_x > Q_c, R_x \in spr[Q_c]\};$$

(observe that any *col* and any *csp* are sets of events, whereas any *spr* is a set of instants).

THEOREM 32 (EXISTENCE OF COLLINEAR SETS OF PARTICLES)

Let $\underset{\sim}{Q},\underset{\sim}{S}$ *be distinct particles with instants* $Q_c \in \underset{\sim}{Q}$, $S_c \in \underset{\sim}{S}$ *such that* $Q_c \simeq S_c$. *Then* $COL[\underset{\sim}{Q},\underset{\sim}{S}]$ *is a collinear set of particles. That is, for any particles* $\underset{\sim}{Q},\underset{\sim}{S},\underset{\sim}{T},\underset{\sim}{U}$ *with* $\underset{\sim}{Q}$ *and* $\underset{\sim}{S}$ *as above,*

$$[\underset{\sim}{Q},\underset{\sim}{S},\underset{\sim}{T}] \text{ and } [\underset{\sim}{Q},\underset{\sim}{S},\underset{\sim}{U}] \Rightarrow [\underset{\sim}{Q},\underset{\sim}{S},\underset{\sim}{T},\underset{\sim}{U}] .$$

This theorem is a consequence of Axioms I (§2.2) and IX (§2.11), and Theorems 4 (§2.7), 9 (§3.2), 14 (§3.6), 27 (§6.1), 28 (§6.1), 29 (§6.2) and 31 (§6.3). It is implicitly used in many of the following theorems.

PROOF. By the Axiom of Incidence (Axiom IX, §2.11) and the previous theorem, there are particles in $COL[\underset{\sim}{Q},\underset{\sim}{S}]$ which do not coincide with $\underset{\sim}{Q}$ and $\underset{\sim}{S}$ at $[Q_c]$; so we must show that, for each instant $Q_z \in \underset{\sim}{Q}$, there are two distinct optical lines, each containing Q_z and one instant from each particle of $COL[\underset{\sim}{Q},\underset{\sim}{S}] \setminus \{\underset{\sim}{Q}\}$, in accordance with the definition preceding this theorem.

Case 1. $Q_z \neq Q_c$

By Theorem 31 (§6.3), there is a particle $\underset{\sim}{T} \in \hat{\underset{\sim}{Q}}_S \subset CSP\langle\underset{\sim}{Q},\underset{\sim}{S}\rangle$ such that $\langle\underset{\sim}{Q},\underset{\sim}{S},\underset{\sim}{T}\rangle$. By Theorem 4 (§2.7) there are two distinct optical lines, each containing Q_z and one instant from each particle of $COL[\underset{\sim}{Q},\underset{\sim}{S}] \setminus \{\underset{\sim}{Q}\}$.

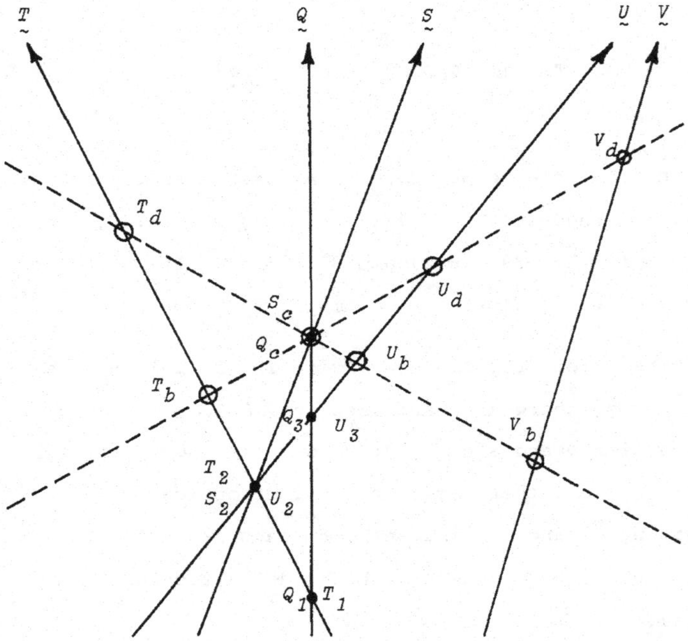

Fig. 33

Case 3. $Q_z = Q_c$ (see Fig. 33)

By Theorem 28 (§6.1) if we consider an instant $Q_1 \in \underset{\sim}{Q}$ with $Q_1 < Q_c$, there is a particle $\underset{\sim}{T}$ with instants $T_1, T_2 \in \underset{\sim}{T}$ and $S_2 \in \underset{\sim}{S}$ such that

$$Q_1 \simeq T_1 < T_2 \simeq S_2 < S_c \simeq Q_c;$$

and there is a particle $\underset{\sim}{U}$ with instants $U_2, U_3 \in \underset{\sim}{U}$ and $Q_3 \in \underset{\sim}{Q}$ such that

$$S_2 \simeq T_2 \simeq U_2 < U_3 \simeq Q_3 < Q_c \simeq S_c.$$

By Theorem 27 (§6.1),

$$csp\langle \underset{\sim}{Q}, \underset{\sim}{T}\rangle \subset csp\langle \underset{\sim}{S}, \underset{\sim}{T}\rangle \subset csp\langle \underset{\sim}{Q}, \underset{\sim}{S}\rangle$$

and

$$csp\langle \underset{\sim}{Q}, \underset{\sim}{T}\rangle \subset csp\langle \underset{\sim}{T}, \underset{\sim}{U}\rangle \subset csp\langle \underset{\sim}{Q}, \underset{\sim}{U}\rangle.$$

Now by Theorem 29 (§6.2), if we define the sides of $\underset{\sim}{Q}$ in $csp\langle \underset{\sim}{Q}, \underset{\sim}{T}\rangle$ such that:

$\underset{\sim}{T}$ is on the left of $\underset{\sim}{Q}$ after $[Q_1]$ in $csp\langle \underset{\sim}{Q}, \underset{\sim}{T}\rangle$, then

$\underset{\sim}{U}$ is on the right of $\underset{\sim}{Q}$ after $[Q_3]$ in $csp\langle \underset{\sim}{Q}, \underset{\sim}{T}\rangle$.

Thus, if we let $T_b \overset{def}{=} f^{-1}_{QT}(Q_c)$, $T_d \overset{def}{=} f_{TQ}(Q_c)$ and $U_b \overset{def}{=} f^{-1}_{QU}(Q_c)$, $U_d \overset{def}{=} f_{UQ}(Q_c)$, we have

(1) $$|T_b, Q_c, U_d\rangle \text{ and } |U_b, Q_c, T_d\rangle.$$

For each particle $\underset{\sim}{V} \in COL[\underset{\sim}{Q},\underset{\sim}{S}] \setminus CSP\langle\underset{\sim}{Q},\underset{\sim}{S}\rangle$, we let $V_b \overset{def}{=} f^{-1}_{QV}(Q_c)$ and $V_d \overset{def}{=} f_{VQ}(Q_c)$; so $V_b < V_d$. Since neither $\underset{\sim}{T}$ nor $\underset{\sim}{V}$ coincides with $\underset{\sim}{Q}$ at the event $[Q_c]$, each of the four instants T_b, T_d, V_b, V_d appears in one (but not two) of the optical lines described in the above Case 1. Now $V_b < Q_c < T_d$ and $T_b < Q_c < V_d$ and since signal functions are one-to-one (by the Signal Axiom (Axiom I, §2.2)), the only possible combinations of signal relations between the four instants T_b, T_d, V_b, V_d are:

(i) $T_b \, \sigma \, V_b$ and $T_d \, \sigma \, V_d$ whence $|T_b, V_b, Q_c\rangle$ and $|Q_c, T_d, V_d\rangle$,

(ii) $V_b \, \sigma \, T_b$ and $T_d \, \sigma \, V_d$ whence $|V_b, T_b, Q_c\rangle$ and $|Q_c, T_d, V_d\rangle$,

(iii) $T_b \, \sigma \, V_b$ and $V_d \, \sigma \, T_d$ whence $|T_b, V_b, Q_c\rangle$ and $|Q_c, V_d, T_d\rangle$,

(iv) $V_b \, \sigma \, T_b$ and $V_d \, \sigma \, T_d$ whence $|V_b, T_b, Q_c\rangle$ and $|Q_c, V_d, T_d\rangle$,

(v) $T_b \, \sigma \, V_d$ and $V_b \, \sigma \, T_d$ whence $|T_b, Q_c, V_d\rangle$ and $|V_b, Q_c, T_d\rangle$.

By (1) and Theorem 4 (§2.7), the relations (i)-(v) imply that the instants V_b, V_d belong (one-to-one) to the uniquely determined optical lines containing T_b, Q_c, U_d and U_b, Q_c, T_d. Now $\underset{\sim}{V}$ was any particle in $COL[\underset{\sim}{Q},\underset{\sim}{S}] \setminus CSP\langle\underset{\sim}{Q},\underset{\sim}{S}\rangle$, so there are two distinct optical lines, each containing Q_c and one instant from each particle of $COL[\underset{\sim}{Q},\underset{\sim}{S}] \setminus CSP\langle\underset{\sim}{Q},\underset{\sim}{S}\rangle$. By Theorem 14 (§3.6), there are two distinct optical lines, each containing Q_c and one instant from each particle of $COL[\underset{\sim}{Q},\underset{\sim}{S}] \setminus \{\underset{\sim}{Q}\}$; the proof is now complete.

The relations (i)-(iv) can not occur, as is easily shown by a continuity argument involving Theorem 9 (§3.2), so the remaining relations (v) apply; that is

$$|T_b, Q_c, V_d\rangle \quad \text{and} \quad |V_b, Q_c, T_d\rangle . \quad \square$$

COROLLARY 1. *If $\underset{\sim}{Q}, \underset{\sim}{R}, \underset{\sim}{S}$ are three distinct particles which coincide at three distinct events, then*

$$COL[\underset{\sim}{Q}, \underset{\sim}{R}] = COL[\underset{\sim}{Q}, \underset{\sim}{S}] = COL[\underset{\sim}{R}, \underset{\sim}{S}] \quad \text{and}$$
$$col[\underset{\sim}{Q}, \underset{\sim}{R}] = col[\underset{\sim}{Q}, \underset{\sim}{S}] = col[\underset{\sim}{R}, \underset{\sim}{S}] .$$

This corollary is a consequence of the above theorem and Theorem 30 (§6.3). It is used in the proof of Theorem 33 (§6.4) and Corollary 2 of Theorem 33 (§6.4).

PROOF. For any $\underset{\sim}{T} \in COL[\underset{\sim}{Q}, \underset{\sim}{R}]$ it follows by definition that $[\underset{\sim}{Q}, \underset{\sim}{R}, \underset{\sim}{T}]$. By Theorem 30 (§6.3) we know that $[\underset{\sim}{Q}, \underset{\sim}{R}, \underset{\sim}{S}]$ and so the above theorem implies that $[\underset{\sim}{Q}, \underset{\sim}{R}, \underset{\sim}{S}, \underset{\sim}{T}]$. That is,

$$\underset{\sim}{T} \in COL[\underset{\sim}{Q}, \underset{\sim}{S}] \text{ and } \underset{\sim}{T} \in COL[\underset{\sim}{R}, \underset{\sim}{S}],$$
$$COL[\underset{\sim}{Q}, \underset{\sim}{R}] \subseteq COL[\underset{\sim}{Q}, \underset{\sim}{S}] \text{ and } COL[\underset{\sim}{Q}, \underset{\sim}{R}] \subseteq COL[\underset{\sim}{R}, \underset{\sim}{S}].$$

By cyclic interchange we obtain the other containment relations, from which the conclusion follows. □

COROLLARY 2. *Let $\underset{\sim}{Q}, \underset{\sim}{R}, \underset{\sim}{S}$ be three distinct particles which coincide at some event. If $\langle \underset{\sim}{Q}, \underset{\sim}{R}, \underset{\sim}{S} \rangle$ after $[Q_c]$, then*

$$COL[\underset{\sim}{Q}, \underset{\sim}{R}] = COL[\underset{\sim}{Q}, \underset{\sim}{S}] = COL[\underset{\sim}{R}, \underset{\sim}{S}] \quad \text{and}$$
$$col[\underset{\sim}{Q}, \underset{\sim}{R}] = col[\underset{\sim}{Q}, \underset{\sim}{S}] = col[\underset{\sim}{R}, \underset{\sim}{S}] .$$

This corollary is a consequence of the above theorem and Corollary 1 of Theorem 30 (§6.3). It is used in the proofs of the next corollary and Theorem 33 (§6.4).

PROOF. The method of proof is the same as for the previous corollary, but with Corollary 1 to Theorem 30 (§6.3) taking the place of Theorem 30. □

COROLLARY 3.

$$COL[\underset{\sim}{Q},\underset{\sim}{S}] = \{\underset{\sim}{U}:\ \underset{\sim}{U} \text{ coincides with two distinct events in } csp\langle\underset{\sim}{Q},\underset{\sim}{S}\rangle,\ \underset{\sim}{U} \in \mathcal{P}\}.$$

This corollary is a consequence of the above theorem, the previous corollary and Theorems 19 (§4.3), 26 (§5.5) and 30 (§6.3). It is used in the proof of Theorem 34 (§6.4).

PROOF. By Theorem 30 (§6.3) any particle which coincides with two events of $csp\langle\underset{\sim}{Q},\underset{\sim}{S}\rangle$ is either a particle of $CSP\langle\underset{\sim}{Q},\underset{\sim}{S}\rangle$, or is collinear with two distinct particles of $CSP\langle\underset{\sim}{Q},\underset{\sim}{S}\rangle$ and therefore with $\underset{\sim}{Q}$ and $\underset{\sim}{S}$, by the previous corollary. That is, the right side of the equation (above) is contained in the left side.

In order to demonstrate the opposite containment, we consider any particle $\underset{\sim}{R} \in COL[\underset{\sim}{Q},\underset{\sim}{S}]$. For any instant $Q_x \in \underset{\sim}{Q}$ with $Q_x > Q_c$, there is by Theorem 26 (§5.5) some particle $\underset{\sim}{T} \in CSP\langle\underset{\sim}{Q},\underset{\sim}{S}\rangle$ $(\subset COL[\underset{\sim}{Q},\underset{\sim}{S}])$ such that

$$\left(f_{QT} \circ f_{TQ}\right)^*(Q_x) = \left(f_{QR} \circ f_{RQ}\right)^*(Q_x) .$$

By Theorem 19 (§4.3), $\underset{\sim}{R}$ coincides with $\underset{\sim}{T}$ at

$$[f^+_{TQ}(Q_x)] \in csp\langle \underset{\sim}{Q}, \underset{\sim}{S}\rangle \subset col[\underset{\sim}{Q}, \underset{\sim}{S}].$$

Similarly by taking any other instant $Q_y \in \underset{\sim}{Q}$ with $Q_c < Q_y \neq Q_x$, we find that $\underset{\sim}{R}$ coincides with another event of $csp\langle \underset{\sim}{Q}, \underset{\sim}{S}\rangle$; this completes the proof. □

THEOREM 33 (Existence of Particles in a Collinear Set)

Let $\underset{\sim}{Q}, \underset{\sim}{S}$ *be distinct particles with instants* $Q_c \in \underset{\sim}{Q}$, $S_c \in \underset{\sim}{S}$ *such that* $Q_c \simeq S_c$; *let the right side of* $\underset{\sim}{Q}$ *in* $col[\underset{\sim}{Q}, \underset{\sim}{S}]$ *be such that* $\underset{\sim}{S}$ *is on the right side of* $\underset{\sim}{Q}$ *after* $[Q_c]$. *Given any four instants* $Q_w, Q_x, Q_y, Q_z \in \underset{\sim}{Q}$ *with* $Q_w < Q_y$ *and* $Q_x < Q_z$, *there is a particle* $\underset{\sim}{R} \in COL[\underset{\sim}{Q}, \underset{\sim}{S}]$ such that

$$\left(f_{QR} \circ f_{RQ}\right)^*(Q_w) = Q_x \text{ and } \left(f_{QR} \circ f_{RQ}\right)^*(Q_y) = Q_z .$$

(see Fig. 34)

REMARK. The particle $\underset{\sim}{S}$ is included in the statement of this theorem so that the sides of $\underset{\sim}{Q}$ in $col[\underset{\sim}{Q}, \underset{\sim}{S}]$ can be specified.

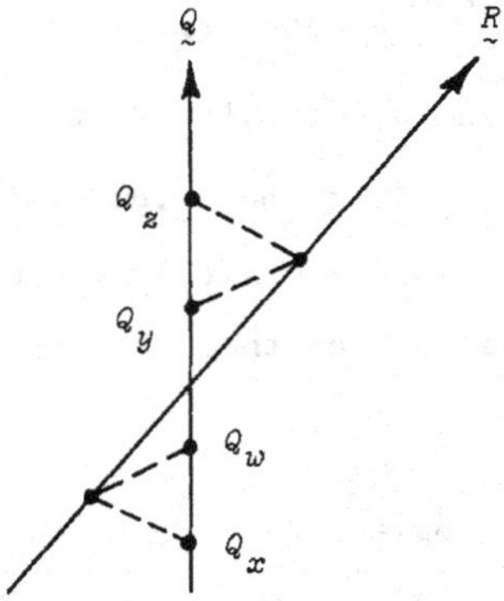

Fig. 34 In this illustration $Q_w > Q_x$ and $Q_y < Q_z$.

This theorem is a consequence of Theorems 6 (§2.9), 17 (§4.2), 18 (§4.3), 19 (§4.3), 26 (§5.5), 28 (§6.1), 29 (§6.2), and Corollaries 1 and 2 of Theorem 32 (§6.4). It is used in the proof of Theorems 34 (§6.4), 35 (§7.1), 36 (§7.1), 45 (§7.4) and 48 (§7.5).

PROOF (see Figs. 35 and 36)

Take any instants $Q_0, Q_2 \in \underset{\sim}{Q}$ with

(1) $$Q_0 < Q_2 < min\{Q_c, Q_w, Q_x, Q_y, Q_z\} .$$

By Theorem 28 (§6.1) there are particles $\underset{\sim}{T}, \underset{\sim}{U}$ and instants $S_3 \in \underset{\sim}{S}$;

$T_1, T_2, T_3 \in \underset{\sim}{T}$ and $U_0, U_1 \in \underset{\sim}{U}$ such that

$$Q_0 \simeq U_0 < U_1 \simeq T_1 < T_2 \simeq Q_2 \text{ and } Q_2 \simeq T_2 < T_3 \simeq S_3 < S_c \simeq Q_c.$$

By Corollary 1 to the previous theorem,

$$(2) \qquad COL[\underset{\sim}{Q},\underset{\sim}{U}] = COL[\underset{\sim}{Q},\underset{\sim}{T}] = COL[\underset{\sim}{Q},\underset{\sim}{S}] \text{ and}$$
$$col[\underset{\sim}{Q},\underset{\sim}{U}] = col[\underset{\sim}{Q},\underset{\sim}{T}] = col[\underset{\sim}{Q},\underset{\sim}{S}].$$

Since $\underset{\sim}{S}$ is on the right side of $\underset{\sim}{Q}$ after $[Q_c]$ in $col[\underset{\sim}{Q},\underset{\sim}{S}]$, that is in $col[\underset{\sim}{Q},\underset{\sim}{T}]$, Theorem 29 (6.2) implies that $\underset{\sim}{S}$ crosses $\underset{\sim}{Q}$ at $[Q_c]$ in $col[\underset{\sim}{Q},\underset{\sim}{T}]$, that is in $col[\underset{\sim}{Q},\underset{\sim}{S}]$. By Theorem 6 (§2.9), $\underset{\sim}{S}$ can only cross $\underset{\sim}{Q}$ at one event, so by Theorem 17 (§4.2), $\underset{\sim}{S}$ is on the left side of $\underset{\sim}{Q}$ before $[Q_c]$ in $col[\underset{\sim}{Q},\underset{\sim}{S}]$. Now the sides of $\underset{\sim}{Q}$ in $col[\underset{\sim}{Q},\underset{\sim}{S}]$ are completely specified.

Theorem 26 (§5.5) implies the existence of particles $\underset{\sim}{V} \in CSP\langle \underset{\sim}{Q},\underset{\sim}{T}\rangle$ and $\underset{\sim}{W} \in CSP\langle \underset{\sim}{Q},\underset{\sim}{U}\rangle$ such that

$$(3) \qquad \left(f_{QV} \circ f_{VQ}\right)^*(Q_w) = Q_x \text{ and } \left(f_{QW} \circ f_{WQ}\right)^*(Q_w) = Q_x.$$

By Theorem 19 (§4.3) $\underset{\sim}{V}$ and $\underset{\sim}{W}$ coincide at the event $[f^+_{VQ}(Q_w)]$. Now $\underset{\sim}{Q},\underset{\sim}{V},\underset{\sim}{W}$ satisfy the conditions of Corollary 1 to the previous theorem and $\underset{\sim}{Q},\underset{\sim}{U},\underset{\sim}{W}$ satisfy the conditions of Corollary 2 to the previous theorem, so by (2),

$$(4) \qquad COL[\underset{\sim}{V},\underset{\sim}{W}] = COL[\underset{\sim}{Q},\underset{\sim}{W}] = COL[\underset{\sim}{Q},\underset{\sim}{U}] = COL[\underset{\sim}{Q},\underset{\sim}{S}].$$

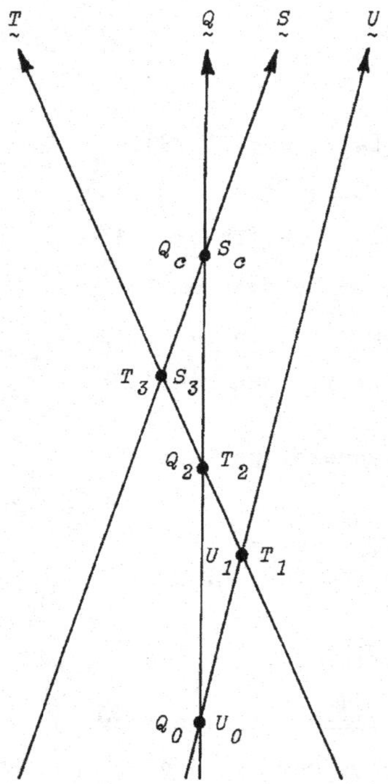

Fig. 35

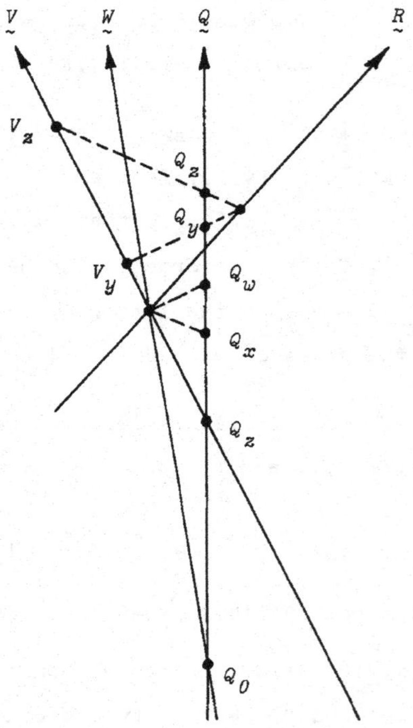

Fig. 36

Now let $V_y \stackrel{def}{=} f^{+}_{VQ}(Q_y)$ and let $V_z \stackrel{def}{=} f^{-}_{VQ}(Q_z)$. By Theorem 26 (§5.5) there is a particle $\underset{\sim}{R} \in CSP\langle \underset{\sim}{V}, \underset{\sim}{W}\rangle$ such that

$$\left(f_{VR} \circ f_{RV}\right)^{*}(V_y) = V_z,$$

and by Theorem 18 (§4.3) and the above definitions it follows that

(5) $$\left(f_{QR} \circ f_{RQ}\right)^{*}(Q_y) = Q_z.$$

Since $\underset{\sim}{R} \in CSP\langle \underset{\sim}{V}, \underset{\sim}{W}\rangle$, equation (3) and the remarks following it imply that

(6) $$\left(f_{QR} \circ f_{RQ}\right)^{*}(Q_w) = Q_x,$$

and equation (4) implies that

(7) $$\underset{\sim}{R} \in COL[\underset{\sim}{Q}, \underset{\sim}{S}],$$

which, together with (5) and (6), completes the proof. □

COROLLARY 1. *Let $\underset{\sim}{Q}, \underset{\sim}{S}$ be distinct particles which coincide at the event $[Q_c]$. Then $\underset{\sim}{S}$ crosses $\underset{\sim}{Q}$ at the event $[Q_c]$ in $col[\underset{\sim}{Q}, \underset{\sim}{S}]$.*

This corollary is used in the proof of Theorems 35 (§7.1) 36 (§7.1), 37 (§7.2) and Corollary 2 of Theorem 58 (§9.3).

PROOF. See the first paragraph of the previous proof. □

COROLLARY 2. *Given a particle* $\underset{\sim}{S}$ *and an event* $[Q_y]$ *which does not coincide with* $\underset{\sim}{S}$*, there is a unique col* which *contains* $[Q_y]$ *and all events coincident with* $\underset{\sim}{S}$.

This corollary is a consequence of Axiom X (§2.12) and Corollary 1 of Theorem 32 (§6.4). It is used in the proof of Theorems 60 (§9.4) and 61 (§9.5).

PROOF. Take an instant $S_x \in \underset{\sim}{S}$ such that

$$(1) \qquad S_x < f^{-1}_{QS}(Q_y) \quad .$$

By the Axiom of Connectedness (Axiom X, §2.12), there are particles $\underset{\sim}{T}, \underset{\sim}{U}$ with instants $T_y, T_z \in \underset{\sim}{T}$ and $U_x, U_z \in \underset{\sim}{U}$ such that

$$(2) \qquad S_x \simeq U_x \quad \text{and} \quad U_z \simeq T_z \quad \text{and} \quad T_y \simeq Q_y \quad .$$

By (1) and (2) and the above theorem, there is a particle $\underset{\sim}{R} \in COL[\underset{\sim}{T}, \underset{\sim}{U}]$ such that $\underset{\sim}{R}$ coincides with both events $[Q_y]$ and $[S_x]$. Now

$$[Q_y] \in col[\underset{\sim}{R}, \underset{\sim}{S}] \text{ and } \underset{\sim}{S} \in COL[\underset{\sim}{R}, \underset{\sim}{S}]$$

Furthermore, Corollary 1 of Theorem 32 (§6.4) implies that $col[\underset{\sim}{R}, \underset{\sim}{S}]$ is independent of the instant S_x. □

COROLLARY 3 (Characterisation of Optical Lines)

If S_x *and* Q_y *are non-coincident instants such that* $S_x \ \sigma \ Q_y$ *, then there is a unique optical line containing* S_x *and* Q_y.

This corollary is used in many of the subsequent theorems.

PROOF. Since $\underset{\sim}{S}$ does not coincide with $\underset{\sim}{Q}$ at $[Q_y]$, the conditions of the previous corollary are satisfied. □

This corollary is a much stronger result than Theorem 4 (§2.7).

THEOREM 34 (Containment Theorem)

Let $\underset{\sim}{Q}, \underset{\sim}{S}$ *be distinct particles which coincide at some event* $[Q_c]$. *Then*

$$COL[\underset{\sim}{Q},\underset{\sim}{S}] = \{\underset{\sim}{U} : \underset{\sim}{U} \text{ coincides with two distinct events of } col[\underset{\sim}{Q},\underset{\sim}{S}],\ \underset{\sim}{U} \in \mathcal{P}\}.$$

REMARK. Any two distinct instants of a particle must be temporally ordered, so there can be no particle coincident with two unordered events.

This theorem is a consequence of Theorem 6 (§2.9), Corollary 3 to Theorem 32 (§6.4) and Theorem 33 (§6.4). It is used implicitly in many of the subsequent theorems.

PROOF. By Corollary 3 to Theorem 32 (§6.4), $COL[\underset{\sim}{Q},\underset{\sim}{S}]$ is contained in the right side of the above equation. In order to demonstrate the opposite containment, consider a particle $\underset{\sim}{U}$ having two instants $U_1, U_2 \in \underset{\sim}{U}$ with $U_1 < U_2$ such that

$$[U_1],\ [U_2] \in col[\underset{\sim}{Q},\underset{\sim}{S}].$$

Let $Q_w \overset{def}{=} \underset{QU}{f^+}(U_1)$, let $Q_x \overset{def}{=} \underset{QU}{f^-}(U_1)$, let $Q_y \overset{def}{=} \underset{QU}{f^+}(U_2)$ and let $Q_z \overset{def}{=} \underset{QU}{f^-}(U_2)$; the conditions of the previous theorem are now satisfied so there is a particle $\underset{\sim}{R} \in COL[\underset{\sim}{Q},\underset{\sim}{S}]$ such that $\underset{\sim}{R}$ coincides with $\underset{\sim}{U}$ at the two distinct events $[U_1]$ and $[U_2]$. By Theorem 6 (§2.9),

$$\underset{\sim}{U} \simeq \underset{\sim}{R} \in COL[\underset{\sim}{Q},\underset{\sim}{S}],$$

which completes the proof. □

CHAPTER 7

THEORY OF PARALLELS

In previous sections we have demonstrated the existence of collinear sets of particles, and we have seen that collinear sets of particles have some properties analogous to coplanar sets of lines in the theory of absolute geometry. Whereas a "parallel postulate" is required to distinguish between the Euclidean and Bolyai-Lobachevskian geometries, no special "parallel postulate" is required in the present treatment. However, until we prove the theorem which I take the liberty of naming the "Euclidean" Parallel Theorem (Theorem 48, §7.5), we must consider the possibility of there being two different types of parallels.

Having shown that there is only one type of parallel, it follows that each particle has a "natural time-scale" which is determined to within an arbitrary increasing linear transformation. Then it is not difficult to show that modified signal functions are linear, and the ensueing discussion of one-dimensional kinematics is taken up in the next chapter.

In most of the subsequent proofs, questions of collinearity are trivial due to the results of §6.4. We let *COL* represent any (maximal) collinear set of particles, and we let *col* represent the corresponding set of instants. Sometimes we shall abbreviate the statements of theorems by not mentioning *COL* or *col* explicitly.

§7.1 Divergent and Convergent Parallels

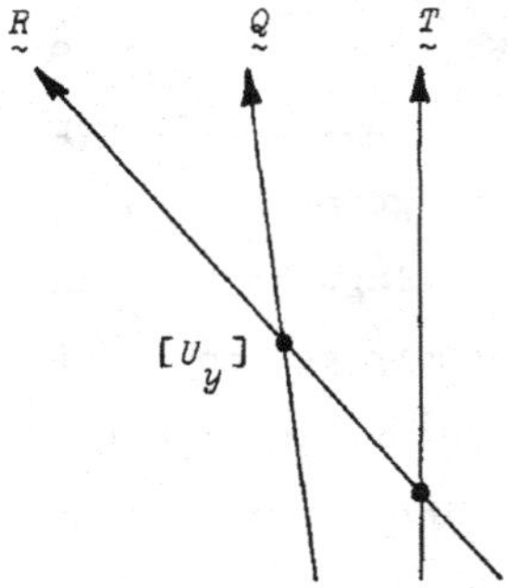

Fig. 37 $\underset{\sim}{Q} \vee (\underset{\sim}{T},[U_y])$

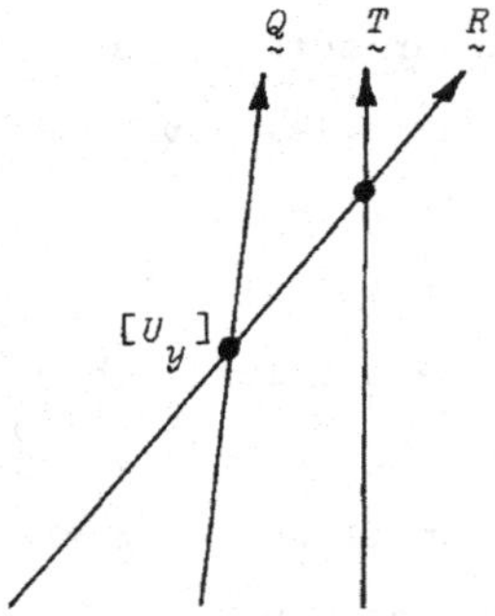

Fig. 38 $\underset{\sim}{Q} \wedge (\underset{\sim}{T},[U_y])$

Given particles $\underset{\sim}{Q},\underset{\sim}{T} \in COL$ and an event $[Q_y] \in col$, we say that $\underset{\sim}{Q}$ is a *divergent parallel* to $\underset{\sim}{T}$ through the event $[U_y]$ (see Fig. 37) and we write $\underset{\sim}{Q} \vee (\underset{\sim}{T},[U_y])$ if:

(i) $\underset{\sim}{Q}$ coincides with the event $[U_y]$,

(ii) $\underset{\sim}{Q}$ does not coincide with $\underset{\sim}{T}$ at any event, and

(iii) for each particle $\underset{\sim}{R} \in SPR[U_y]$ such that

$$\langle \underset{\sim}{R},\underset{\sim}{Q},\underset{\sim}{T} \rangle \; after \; [U_y] \text{ and } \underset{\sim}{R} \neq \underset{\sim}{Q} ,$$

$\underset{\sim}{R}$ coincides with $\underset{\sim}{T}$ at some event before $[U_y]$.

Sometimes we merely say that $\underset{\sim}{Q}$ *diverges from* $\underset{\sim}{T}$ through $[U_y]$.

Similarly, we say that $\underset{\sim}{Q}$ is a *convergent parallel* to $\underset{\sim}{T}$ through the event $[U_y]$ (see Fig. 38) and we write $\underset{\sim}{Q} \wedge (\underset{\sim}{T}, [U_y])$ if:

(i) $\underset{\sim}{Q}$ coincides with the event $[U_y]$

(ii) $\underset{\sim}{Q}$ does not coincide with $\underset{\sim}{T}$ at any event, and

(iii) for each particle $\underset{\sim}{R} \in SPR[U_y]$ such that

$$\langle \underset{\sim}{R}, \underset{\sim}{Q}, \underset{\sim}{T} \rangle \textit{ before } [U_y] \text{ and } \underset{\sim}{R} \neq \underset{\sim}{Q} ,$$

$\underset{\sim}{R}$ coincides with $\underset{\sim}{T}$ at some event after $[U_y]$.
Sometimes we merely say that $\underset{\sim}{Q}$ *converges to* $\underset{\sim}{T}$ through $[U_y]$.

In the remainder of this section we will often use the symbol $\|$ to represent either $\vee$ or $\wedge$, where it is implied that the substitution is consistent in any statement or proof.

We now show that *parallelism is a relation between particles* by proving the following:

THEOREM 35 (Transmissibility of Parallelism)
If $\underset{\sim}{Q} \parallel (\underset{\sim}{T}, [U_y])$ *and* $Q_c \in \underset{\sim}{Q}$, *then* $\underset{\sim}{Q} \parallel (\underset{\sim}{T}, [Q_c])$.
That is, $\underset{\sim}{Q}$ *is parallel to* $\underset{\sim}{T}$ *and we write* $\underset{\sim}{Q} \vee \underset{\sim}{T}$ *or* $\underset{\sim}{Q} \wedge \underset{\sim}{T}$, *as the case may be; or simply* $\underset{\sim}{Q} \parallel \underset{\sim}{T}$ *with the above convention.*

This theorem is a consequence of Theorem 33 (§6.4) and Corollary 1 of Theorem 33 (§6.4). It is used in the proof of Theorems 36 (§7.1), 37 (§7.2), 38 (§7.2), 40 (§7.3) and 42 (§7.3).

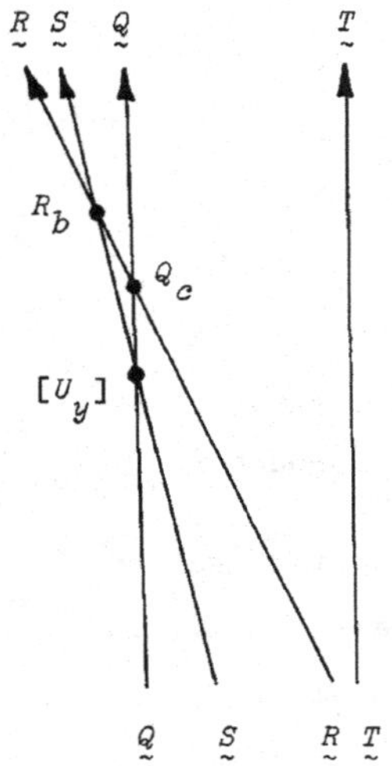

Fig. 39

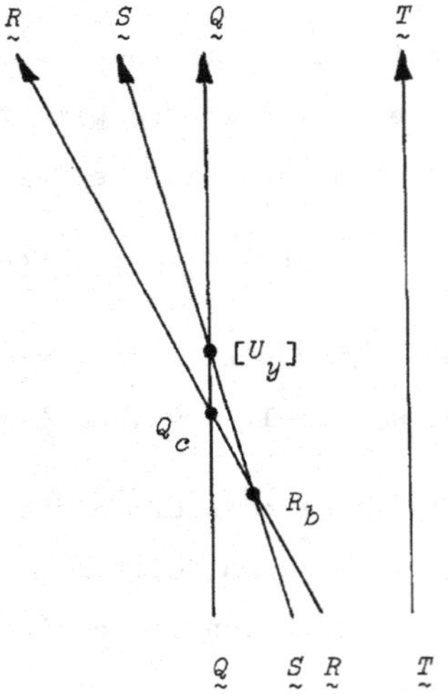

Fig. 40

PROOF. *(a) Transmissibility of Divergent Parallelism* Suppose the contrary; that is, suppose there is a particle $\underset{\sim}{R}$ which coincides with $\underset{\sim}{Q}$ at $[Q_c]$ such that $\langle \underset{\sim}{Q}, \underset{\sim}{R}, \underset{\sim}{T} \rangle$ *before* $[Q_c]$ and $\underset{\sim}{R} \nparallel \underset{\sim}{Q}$. Take an instant $R_b \in \underset{\sim}{R}$ with $R_b > Q_c$ if $Q_c > U_y$ (see Fig. 39) or with $R_b < Q_c$ if $Q_c < U_y$ (see Fig. 40). By Theorem 33 (§6.4) there is a particle $\underset{\sim}{S}$ which coincides with

$\underset{\sim}{Q}$ at $[U_y]$ and with $\underset{\sim}{R}$ at $[R_b]$. By Corollary 1 of Theorem 33 (§6.4)

$$<\underset{\sim}{Q},\underset{\sim}{S},\underset{\sim}{R},\underset{\sim}{T}> \textit{ before } \min\{[U_y],[R_b]\}$$

so $\underset{\sim}{S}$ does not coincide with any event of $\underset{\sim}{T}$ before $[U_y]$; but this contradicts the third requirement in the definition of $\underset{\sim}{Q}$.

(b) Transmissibility of Convergent Parallelism

The proof is similar to the proof of (a) with the expressions "before" and "min" changed to "after" and "max", respectively. □

If we want to specify that $\underset{\sim}{Q}$ diverges from $\underset{\sim}{T}$, or that $\underset{\sim}{Q}$ converges to $\underset{\sim}{T}$, or that $\underset{\sim}{Q}$ is parallel to $\underset{\sim}{T}$, we use the more concise expressions $\underset{\sim}{Q} \vee \underset{\sim}{T}$, $\underset{\sim}{Q} \wedge \underset{\sim}{T}$, and $\underset{\sim}{Q} \parallel \underset{\sim}{T}$, respectively. At this stage we have not shown that parallel particles exist or that the relations of parallelism are symmetric. For convenience we define both relations of parallelism to be reflexive, so that each particle is (trivially) parallel to itself, in both the divergent and the convergent sense. We extend the definitions of parallelism to apply to observers, so that: $\hat{\underset{\sim}{Q}} \parallel \hat{\underset{\sim}{T}} \Leftrightarrow$ for all $\underset{\sim}{R} \in \hat{\underset{\sim}{Q}}$ and for all $\underset{\sim}{S} \in \hat{\underset{\sim}{T}}$, $\underset{\sim}{R} \parallel \underset{\sim}{S}$. We now extend the definition of mid-way and reflected particles (§5.3) so as to apply to particles, such as parallel particles, which need not coincide at any event. As a result of Theorem 31 (§6.3) we can extend the definitions of §5.3 to become: if $\underset{\sim}{Q},\underset{\sim}{S},\underset{\sim}{U}$ are particles such that

$$<\underset{\sim}{Q},\underset{\sim}{S},\underset{\sim}{U}> \text{ and } f_{SQ} \circ f_{QS} = f_{SU} \circ f_{US} ,$$

we say that $\underset{\sim}{S}$ is *mid-way between* $\underset{\sim}{Q}$ and $\underset{\sim}{U}$, and we say that $\underset{\sim}{Q}$ is a *reflection* of $\underset{\sim}{U}$ in $\underset{\sim}{S}$, and that $\underset{\sim}{U}$ is a *reflection* of $\underset{\sim}{Q}$ in $\underset{\sim}{S}$. We also define *reflected events*, so that if $[T_x]$ and $[V_y]$ are events on opposite sides of $\underset{\sim}{S}$ in *col* and if there are instants $S_w, S_z \in \underset{\sim}{S}$ such that

$$[S_w] \; \sigma \; [T_x] \; \sigma \; [S_z] \text{ and } [S_w] \; \sigma \; [V_y] \; \sigma \; [S_z],$$

we say that $[T_x]$ and $[V_y]$ are *reflections* of each other in $\underset{\sim}{S}$, and we write $[T_x] = [V_y]_S$ and $[V_y] = [T_x]_S$. It follows from Theorem 33 (§6.4) and Theorem 19 (§4.3), that each event (of *col* has a unique reflection in each particle (of *COL*). We will now demonstrate the existence of parallel particles of both types and their reflections.

THEOREM 36 (Existence of Parallels and their Reflections)

Let $\underset{\sim}{S}$ *be a particle in COL, and let* $[V_0]$ *be an event in col. There are particles* $\underset{\sim}{Q}, \underset{\sim}{U} \in$ *COL such that*

(i) $\underset{\sim}{U} \parallel (\underset{\sim}{S}, [V_0])$ *and* $\underset{\sim}{Q} \parallel (\underset{\sim}{S}, [V_0]_S)$ *and*

(ii) $\underset{\sim}{U} \in \hat{\underset{\sim}{Q}}_S$ *and* $\underset{\sim}{Q} \in \hat{\underset{\sim}{U}}_S$.

This theorem is a consequence of Theorems 6 (§2.9), 9 (§3.2), 17 (§4.2), 19 (§4.3), 21 (§5.1), 24 (§5.3), 33 (§6.4) and Corollary 1 of 33 (§6.4), and 35 (§7.1). It is used in the proof of Theorems 37 (§7.2), 39 (§7.3), 40 (§7.3), 41(§7.3) and its corollary, Theorem 43 (§7.3) and its corollary, and Theorems 46 (§7.5) and 61 (§9.5).

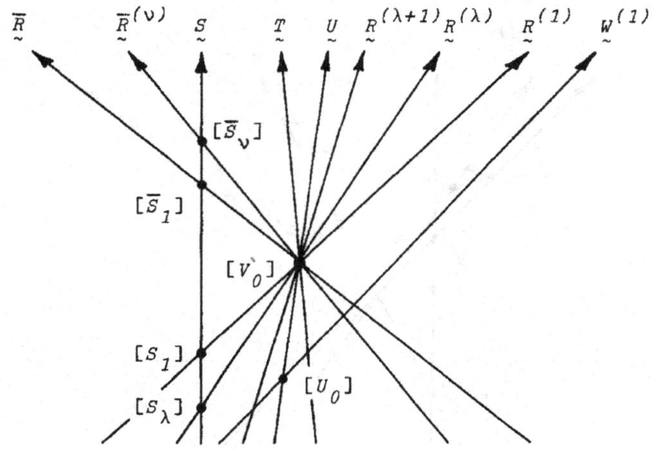

Fig. 41

PROOF. We define the sides of $\underset{\sim}{S}$ so that $[V_0]$ is on the right side of $\underset{\sim}{S}$ in col.

Case (a)   Divergent Parallels

(i) We first show that there is a particle $\underset{\sim}{U}$ γ $(\underset{\sim}{S},[V_0])$. Let $(S_\lambda\colon \lambda=1,2,\cdots;\ S_\lambda\ \varepsilon\ \underset{\sim}{S})$ be an unbounded decreasing sequence of instants with $S_1 < V_0$. By Theorem 33 (§6.4) for each positive integer λ there is a particle $\underset{\sim}{R}^{(\lambda)}$ which coincides with the events $[V_0]$ and $[S_\lambda]$, and which is contained in COL (see Fig. 41).

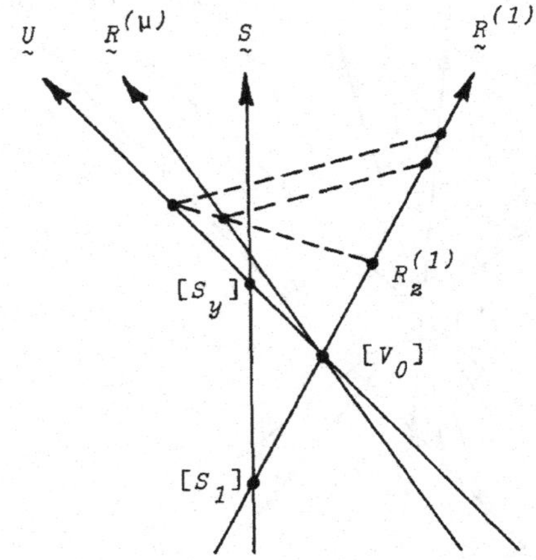

Fig. 42

By the corollary to Theorem 33 (§6.4), for each λ , $\langle \underset{\sim}{S}, \underset{\sim}{R}^{(\lambda+1)}, \underset{\sim}{R}^{(\lambda)} \rangle$ *after* $[V_0]$. By Theorem 21 (§5.1) the set $\{\underset{\sim}{R}^{(\lambda)}: \lambda=1,2,\cdots\}$ has a limit particle $\underset{\sim}{U} \in CSP\langle \underset{\sim}{R}^{(1)}, \underset{\sim}{R}^{(2)} \rangle \subset COL$ such that, for any instant $R_z^{(1)} \in \underset{\sim}{R}^{(1)}$ with $R_z^{(1)} > V_0$,

$$(1) \quad f_{R^{(1)}U} \circ f_{UR^{(1)}}(R_z^{(1)}) = \sup_{\lambda}\left\{ f_{R^{(1)}R^{(\lambda)}} \circ f_{R^{(\lambda)}R^{(1)}}(R_z^{(1)})\right\}, \text{ and}$$

for each positive integer λ, $\langle \underset{\sim}{U}, \underset{\sim}{R}^{(\lambda)}, \underset{\sim}{R}^{1} \rangle$. So for each positive integer λ, the limit particle $\underset{\sim}{U}$ is to the right of the particle

$\underset{\sim}{R}^{(\lambda)}$ before $[V_0]$, so $\underset{\sim}{U}$ does not coincide with $\underset{\sim}{S}$ at any event before $[V_0]$.

Next we show that $\underset{\sim}{U}$ does not coincide with $\underset{\sim}{S}$ at any event after $[V_0]$. Suppose the contrary; that is, suppose that $\underset{\sim}{U}$ coincides with $\underset{\sim}{S}$ at some event $[S_y] > [V_0]$ (where $S_y \in \underset{\sim}{S}$). By Corollary 1 of Theorem 33 (§6.4), $\underset{\sim}{U}$ crosses $\underset{\sim}{S}$ at $[S_y]$ (see Fig. 42) so by Theorem 19 (§4.3), for any instant $R_z^{(1)} \in \underset{\sim}{R}^{(1)}$ with $R_z^{(1)} \nparallel S_y$,

$$f_{R^{(1)}U} \circ f_{UR^{(1)}}(R_z^{(1)}) > f_{R^{(1)}S} \circ f_{SR^{(1)}}(R_z^{(1)}).$$

By equation (1), for any particular instant $R_z^{(1)}$, there is a particle $\underset{\sim}{R}^{(\mu)}$ such that

$$f_{R^{(1)}R^{(\mu)}} \circ f_{R^{(\mu)}R^{(1)}}(R_z^{(1)}) > f_{R^{(1)}S} \circ f_{SR^{(1)}}(R_z^{(1)}).$$

By Theorem 19 (§4.3) the events $[V_0]$ and $[f_{R^{(\mu)}R^{(1)}}(R_z^{(1)})]$ are on opposite sides of $\underset{\sim}{S}$, so by Theorem 17 (§4.2) the particle $\underset{\sim}{R}^{(\mu)}$ coincides with $\underset{\sim}{S}$ at some event between $[V_0]$ and $[f_{R^{(\mu)}R^{(1)}}(R_z^{(1)})]$. But $\underset{\sim}{R}^{(\mu)}$ also coincides with $\underset{\sim}{S}$ at $[S_\mu] < [V_0]$, so by Theorem 6 (§2.9) $\underset{\sim}{R}^{(\mu)} \simeq \underset{\sim}{S}$, which is a contradiction.

In order to show that the third condition of the definition (§7.1) is satisfied, consider any $\underset{\sim}{R} \in SPR[V_0]$ such that $\langle \underset{\sim}{S}, \underset{\sim}{U}, \underset{\sim}{R} \rangle$ *after* $[V_0]$ and $\underset{\sim}{R} \nparallel \underset{\sim}{U}$ (see Fig. 43); then there is some positive integer p such that $\langle \underset{\sim}{S}, \underset{\sim}{U}, \underset{\sim}{R}^{(p)}, \underset{\sim}{R} \rangle$ *after* $[V_0]$, and so $\underset{\sim}{R}$ coincides with $\underset{\sim}{S}$ at some event after $[S_p]$ and before $[V_0]$. We have now shown that

$$\underset{\sim}{U} \vee (\underset{\sim}{S}, [V_0]) \ .$$

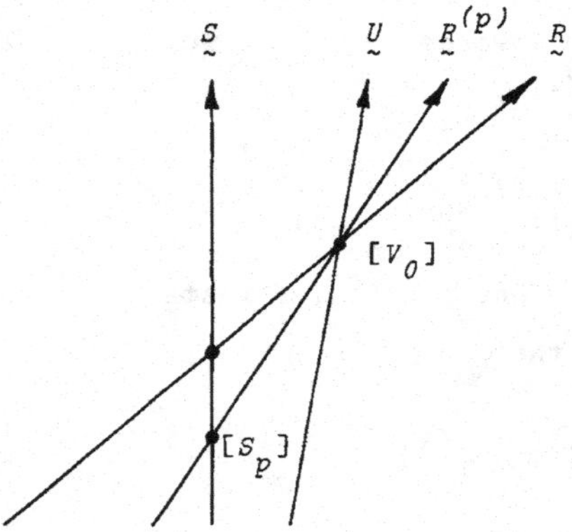

Fig. 43

By Theorem 24 (§5.3) there is a set of particles $\{\underset{\sim}{R}_S^{(\lambda)} : \lambda = 1, 2, \cdots\}$ such that $\underset{\sim}{R}_S^{(\lambda)}$ is a reflection of $\underset{\sim}{R}^{(\lambda)}$ in $\underset{\sim}{S}$. Let $\underset{\sim}{Q}$ be the limit particle of the sequence $\underset{\sim}{R}_S^{(\lambda)} : \lambda = 1, 2, \cdots\}$; then as above,

$$\underset{\sim}{Q} \vee (\underset{\sim}{S}, [V_0]_T) \ .$$

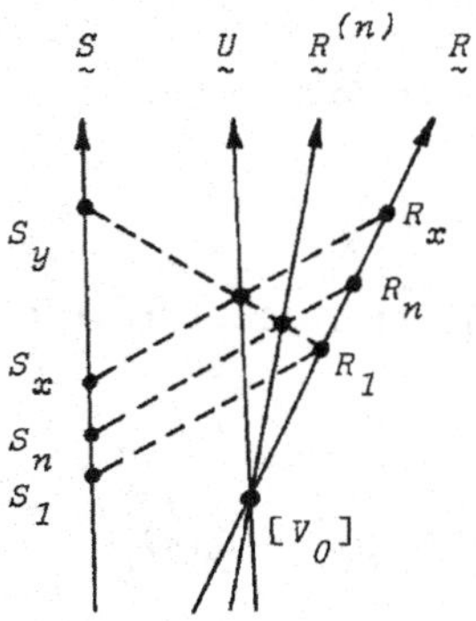

Fig. 44

(ii) We will now show that $\underset{\sim}{Q}$ and $\underset{\sim}{U}$ are reflections of each other in $\underset{\sim}{S}$. We will use the symbol $\underset{\sim}{R}$ as an abbreviation for $\underset{\sim}{R}^{(1)}$. Consider any instant $S_y \in \underset{\sim}{S}$ with $S_y > V_0$. We now define (see Fig. 44):

(2) $$S'_x \overset{def}{=} f^{-1}_{QS} \circ f^{-1}_{SQ}(S_y), \quad S_x \overset{def}{=} f^{-1}_{US} \circ f^{-1}_{SU}(S_y),$$

(3) $$S'_n \overset{def}{=} f^{-1}_{R^{(n)}_S S} \circ f^{-1}_{SR^{(n)}_S}(S_y) = S_n \overset{def}{=} f^{-1}_{R^{(n)} S} \circ f^{-1}_{SR^{(n)}}(S_y)$$

(the set of instants $\{S_n: n=1,2,\cdots\}$ should not be confused with the sequence of instants $(S_\lambda: \lambda=1,2,\cdots)$ of part (i) above),

(4) $$R_1 \overset{def}{=} f^{-1}_{SR}(S_y)$$

(5) $$R_n \overset{def}{=} f_{RR^{(n)}} \circ f_{R^{(n)}R}(R_1), \text{ and}$$

(6) $$R_x \overset{def}{=} f_{RU} \circ f_{UR}(R_1) .$$

By equation (1) (of the proof of (i) above)

(7) $$R_x = \sup_n\{R_n\} .$$

Since $<\underset{\sim}{S}, \underset{\sim}{U}, \underset{\sim}{R}^{(n)}, \underset{\sim}{R}>$ *after* $[V_0]$, it follows that for instants after V_0,

(8) $$f_{SR} = f_{SU} \circ f_{UR} = f_{SR^{(n)}} \circ f_{R^{(n)}R} \quad \text{and}$$

(9) $$f^{-1}_{RS} = f^{-1}_{US} \circ f^{-1}_{RU} = f^{-1}_{R^{(n)}S} \circ f^{-1}_{RR^{(n)}} .$$

Thus $$S_x = f^{-1}_{US} \circ f^{-1}_{SU} \circ f_{SR}(R_1), \text{ by (2) and (4)},$$

$$= f^{-1}_{US} \circ f_{UR}(R_1), \text{ by (8)},$$

$$= f^{-1}_{US} \circ f^{-1}_{RU}(R_x), \text{by (6)},$$

$$= f^{-1}_{RS}(R_x), \text{ by (9)},$$

$$= f^{-1}_{RS}(\sup_n\{R_n\}), \text{ by (7)},$$

$$= \sup_n\left\{f^{-1}_{RS}(R_n)\right\}, \text{ by Theorem 9 (§3.2)},$$

$$= \sup_n \left\{ f^{-1}_{RS} \circ f_{RR^{(n)}} \circ f_{R^{(n)}R}(R_1) \right\}, \text{ by (5)},$$

$$= \sup_n \left\{ f^{-1}_{RS} \circ f_{RR^{(n)}} \circ f_{R^{(n)}R} \circ f^{-1}_{SR}(S_y) \right\}, \text{ by (4)},$$

$$= \sup_n \left\{ f^{-1}_{R^{(n)}S} \circ f^{-1}_{SR^{(n)}}(S_y) \right\}, \text{ by (8) and (9)},$$

$$= \sup_n \{S_n\}, \text{ by (3)}.$$

Now equations analogous to equations (4)-(9) can be defined for the particles $\underset{\sim}{Q}, \underset{\sim}{R}^{(n)}_S$ in place of $\underset{\sim}{U}, \underset{\sim}{R}^{(n)}$ which leads in the same way to

(10) $$S'_x = \sup_n \{S'_n\} = \sup_n \{S_n\} = S_x, \text{ by (3)}.$$

That is, for any $S_y > V_0$,

$$f^{-1}_{US} \circ f^{-1}_{SU}(S_y) = f^{-1}_{QS} \circ f^{-1}_{SQ}(S_y),$$

or equivalently, for any $S_x \nleq V_0$,

(11) $$f_{SU} \circ f_{US}(S_x) = f_{SQ} \circ f_{QS}(S_x).$$

Similarly if we take any instant $U_0 \in \underset{\sim}{U}$ with $U_0 < V_0$, then as in the proof above there are sets of particles $\{\underset{\sim}{W}^{(\lambda)}: \lambda = 1, 2, \cdots\}$ and $\{\underset{\sim}{W}^{(\lambda)}_S: \lambda = 1, 2, \cdots\}$ which have limit particles which are parallel to $\underset{\sim}{S}$ (see Fig. 41). These parallels coincide with the events $[U_0]$ and $[U_0]_S$ respectively, and satisfy an equation similar to equation (11) for all instants $S_x \in \underset{\sim}{S}$ with $S_x \nmid U_0$. By the previous theorem, the parallel which coincides with $[U_0]$ must be $\underset{\sim}{U}$, and by the equation analogous to (11), the other parallel must be $\underset{\sim}{Q}$. That is, for any $S_x \in \underset{\sim}{S}$ with $S_x \nmid U_0$,

$$(12) \qquad f_{SU} \circ f_{US}(S_x) = f_{SQ} \circ f_{QS}(S_x).$$

But U_0 is arbitrary, so (12) applies for all $S_x \in \underset{\sim}{S}$; that is, $\underset{\sim}{Q}$ and $\underset{\sim}{U}$ are reflections of each other in $\underset{\sim}{S}$. This completes the proof for the case of divergent parallels.

Case (b) Convergent Parallels

(i) We first show that there is a particle $\underset{\sim}{T} \wedge (\underset{\sim}{S}, [V_x])$. Let $\{\overline{S}_\nu : \nu = 1, 2, \cdots ; \overline{S}_\nu \in \underset{\sim}{S}\}$ be an unbounded increasing sequence of instants with $\overline{S}_1 > V_0$. (The bars are intended to distinguish the sequence $(\overline{S}_\nu)$ from the previous sequence (S_λ) and do not represent ideal instants as in Chapter 3). As in the proof for the case of divergent parallels (see Fig. 41), for each positive integer ν there is a particle $\underset{\sim}{\overline{R}}^{(\nu)}$ which coincides with the events $[V_0]$ and $[\overline{S}_\nu]$. So for all positive integers λ and ν, $\langle \underset{\sim}{\overline{R}}^{(1)}, \underset{\sim}{\overline{R}}^{(\nu)}, \underset{\sim}{R}^{(\lambda)}, \underset{\sim}{R}^{(1)} \rangle$ and the set $\{\underset{\sim}{\overline{R}}^{(\nu)} : \nu = 1, 2, \cdots\}$ has a limit particle $\underset{\sim}{T}$ such that $\langle \underset{\sim}{\overline{R}}^{(1)}, \underset{\sim}{\overline{R}}^{(\nu)}, \underset{\sim}{T}, \underset{\sim}{U}, \underset{\sim}{R}^{(\lambda)}, \underset{\sim}{R}^{(1)} \rangle$ where $\underset{\sim}{U}$ is the divergent parallel of the previous Case *a*. Hence $\underset{\sim}{T}$ can not coincide with $\underset{\sim}{S}$ at any event after $[V_0]$ and by the above ordering (with respect to $\underset{\sim}{U}$ and $\underset{\sim}{R}^{(1)}$), $\underset{\sim}{T}$ can not coincide with $\underset{\sim}{S}$ at any event before $[V_0]$. In order to show that the third condition of the definition (§ 7.1) is satisfied, we could use an argument analogous to the corresponding argument for divergent parallels above. In the remainder of this proof we have no further use for the symbol $\underset{\sim}{T}$ or for the divergent parallel $\underset{\sim}{U}$ of the above Case a. So we let the symbol $\underset{\sim}{U}$ represent the convergent parallel which has so far been called $\underset{\sim}{T}$; that is, we define the particle $\underset{\sim}{U}$ so that

$$\underset{\sim}{U} \wedge (\underset{\sim}{S}, [V_0])$$

Similarly, as in the previous case, there exists a particle $\underset{\sim}{Q}$ such that

$$\underset{\sim}{Q} \wedge (\underset{\sim}{S}, [V_0]_S].$$

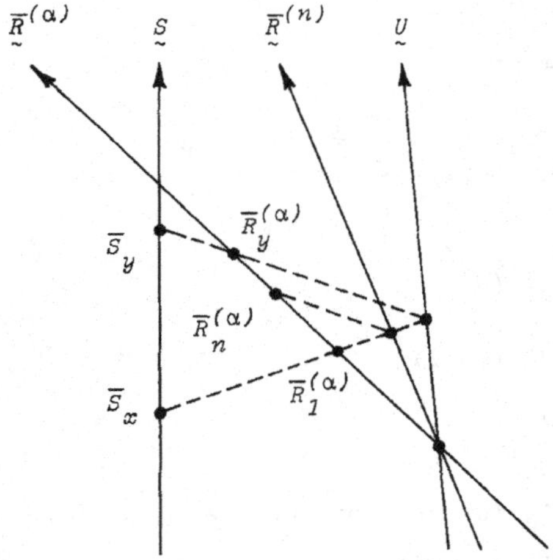

Fig. 45

(ii) We will now show that $\underset{\sim}{Q}$ and $\underset{\sim}{U}$ are reflections of each other in $\underset{\sim}{S}$. The proof is similar to the proof of Case *a*, and is shown in outline (see Fig. 45). The symbols used for this proof are similar to those which were used for the proof of Case *a*, but here they have different meanings. Consider any instant $\overline{S}_x \in \underset{\sim}{S}$ with $\overline{S}_x \nleq V_0$; then there is some positive integer α such that $\overline{S}_\alpha > f_{SU} \circ f_{US}(\overline{S}_x)$. We now define:

(13) $\overline{S}'_y \overset{def}{=} f_{SQ} \circ f_{QS}(\overline{S}_x)$, $\qquad \overline{S}_y \overset{def}{=} f_{SU} \circ f_{US}(\overline{S}_x)$,

(14) $\overline{S}'_n \overset{def}{=} f_{S\overline{R}^{(n)}_S} \circ f_{\overline{R}^{(n)}_S S}(\overline{S}_x) = \overline{S}_n \overset{def}{=} f_{S\overline{R}^{(n)}} \circ f_{\overline{R}^{(n)} S}(\overline{S}_x)$,

(15) $\overline{R}^{(\alpha)}_1 \overset{def}{=} f_{\overline{R}^{(\alpha)} S}(\overline{S}_x)$,

(16) $\overline{R}^{(\alpha)}_n \overset{def}{=} f_{\overline{R}^{(\alpha)}\overline{R}^{(n)}} \circ f_{\overline{R}^{(n)}\overline{R}^{(\alpha)}}(\overline{R}^{(\alpha)}_1)$, and

(17) $\overline{R}^{(\alpha)}_y \overset{def}{=} f_{\overline{R}^{(\alpha)} U} \circ f_{U\overline{R}^{(\alpha)}}(\overline{R}^{(\alpha)}_1)$.

Therefore by Theorem 9 (3.2)

(18) $\overline{R}^{(\alpha)}_y = \sup_n \{\overline{R}^{(\alpha)}_n\}$.

By the choice of $\underset{\sim}{\overline{R}}^{(\alpha)}$, for any integer $n > \alpha$ and for any instants $S_z \in \underset{\sim}{S}$ and $U_z \in \underset{\sim}{U}$ with

$$\overline{S}_x < S_z < \overline{S}_y \text{ and } V_0 < U_z < \overline{S}_y,$$

$$(19)\quad f_{US}(S_z) = f_{U\overline{R}^{(n)}} \circ f_{\overline{R}^{(n)}\overline{R}^{(\alpha)}} \circ f_{\overline{R}^{(\alpha)}S}(S_z) = f_{U\overline{R}^{(n)}} \circ f_{\overline{R}^{(n)}S}(S_z),$$

and

$$(20)\quad f_{SU}(U_z) = f_{S\overline{R}^{(\alpha)}} \circ f_{\overline{R}^{(\alpha)}\overline{R}^{(n)}} \circ f_{\overline{R}^{(n)}U}(U_z) = f_{S\overline{R}^{(n)}} \circ f_{\overline{R}^{(n)}U}(U_z).$$

So $\overline{S}_y = f_{SU} \circ f_{US}(\overline{S}_x)$

$= f_{S\overline{R}^{(\alpha)}} \circ f_{\overline{R}^{(\alpha)}U} \circ f_{U\overline{R}^{(\alpha)}} \circ f_{\overline{R}^{(\alpha)}S}(\overline{S}_x)$

$= f_{S\overline{R}^{(\alpha)}}(sup\{\overline{R}_n^{(\alpha)}\})$, by (15), (17), (18),

$= sup\left\{ f_{S\overline{R}^{(\alpha)}}(\overline{R}_n^{(\alpha)})\right\}$, by Theorem 9 (§3.2),

$= sup\left\{ f_{S\overline{R}^{(\alpha)}} \circ f_{\overline{R}^{(\alpha)}\overline{R}^{(n)}} \circ f_{\overline{R}^{(n)}\overline{R}^{(\alpha)}} \circ f_{\overline{R}^{(\alpha)}S}(\overline{S}_x)\right\}$,

by (15) and (16),

$= sup\left\{ f_{S\overline{R}^{(n)}} \circ f_{\overline{R}^{(n)}S}(\overline{S}_x)\right\}$, by (19), (20),

$= sup\{\overline{S}_n\}$, by (14).

The remainder of the argument is similar to the argument for divergent parallels. □

§7.2 The Parallel Relations are Equivalence Relations

We now show that both relations of parallelism are equivalence relations. It then follows that in any collinear set of particles, there are equivalence classes of parallel particles, of both the divergent and the convergent type.

THEOREM 37 (Symmetry of Parallelism)

Let $\underset{\sim}{Q}, \underset{\sim}{S} \in COL$. If $\underset{\sim}{Q} \parallel \underset{\sim}{S}$ then $\underset{\sim}{S} \parallel \underset{\sim}{Q}$.

This theorem is a consequence of Theorems 19 (§4.3), 23 (§5.3), Corollary 1 of Theorem 33 (§6.4) and Theorems 35 (§7.1) and 36 (§7.1). It is used in the proof of Theorem 38 (§7.2) and implicitly in many subsequent theorems.

PROOF. We have already specified that the relation(s) of parallelism are reflexive (in the remarks following Theorem 35 (§7.1)), so we consider distinct particles $\underset{\sim}{Q}$ and $\underset{\sim}{S}$.

Case (a) Divergent Parallels (See Fig. 46)

We suppose that $\underset{\sim}{S} \not\downarrow \underset{\sim}{Q}$; that is, we suppose that for some $S_c \in \underset{\sim}{S}$, there is a particle $\underset{\sim}{U}$ such that

(1) $$\underset{\sim}{U} \vee (\underset{\sim}{Q}, [S_c]) \text{ and } \underset{\sim}{U} \nmid \underset{\sim}{S}.$$

By Theorem 23 (§5.3) there is a particle $\underset{\sim}{T}$ mid-way between $\underset{\sim}{S}$ and $\underset{\sim}{U}$ and a reflection mapping ϕ such that

$$\phi(\underset{\sim}{T}) \simeq \underset{\sim}{T}, \; \phi(\underset{\sim}{S}) \simeq \underset{\sim}{U}, \; \phi(\underset{\sim}{U}) \simeq \underset{\sim}{S}.$$

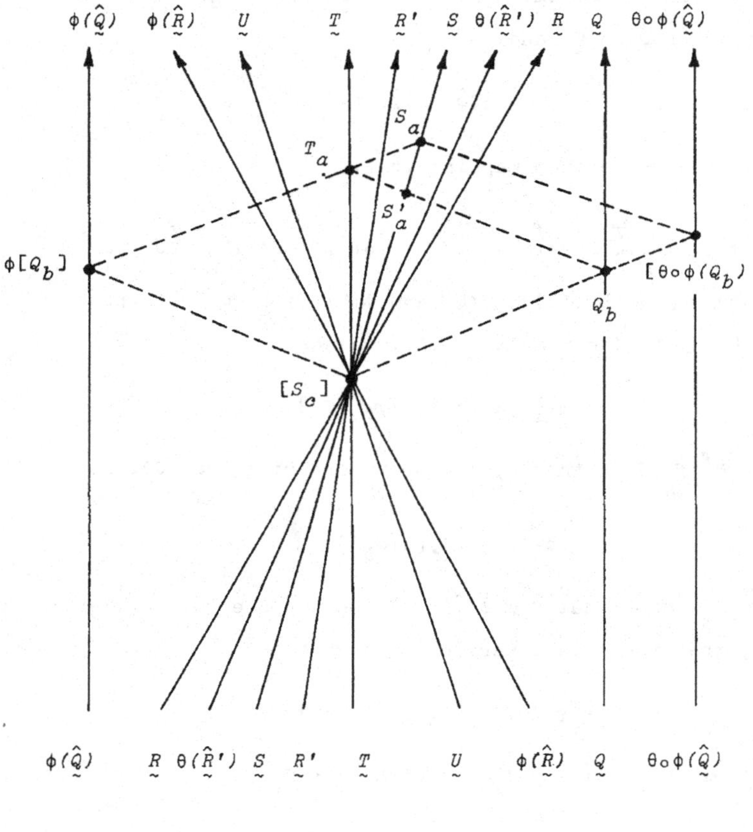

Fig. 46

Now $<\underset{\sim}{S},\underset{\sim}{T},\underset{\sim}{U},\underset{\sim}{Q}>$ *before* $[S_c]$ so

(2) $$\underset{\sim}{Q} \vee \underset{\sim}{S},\ \underset{\sim}{Q} \vee \underset{\sim}{T}, \text{ and } \underset{\sim}{Q} \vee \underset{\sim}{U}.$$

By the previous theorem, there is an observer $\phi(\underset{\sim}{\hat{Q}})$ which is a reflection of $\underset{\sim}{\hat{Q}}$ in $\underset{\sim}{T}$, and

(3) $$\phi(\underset{\sim}{\hat{Q}}) \vee \underset{\sim}{\hat{T}}.$$

Since ϕ is a reflection mapping in $\underset{\sim}{T}$,

$$f_{T\phi(\underset{\sim}{Q})} \circ f_{\phi(\underset{\sim}{Q})T} = f_{TQ} \circ f_{QT} \text{ and } f_{T\phi(U)} \circ f_{\phi(U)T} = f_{TU} \circ f_{UT},$$

and since $\underset{\sim}{U} \vee \underset{\sim}{Q}$ (by (1)), $\phi(\underset{\sim}{Q})$ and $\phi(\underset{\sim}{U})$ $(\simeq \underset{\sim}{S})$ do not cross before $[S_c]$, by Theorem 19 (§4.3). Hence

$$<\phi(\underset{\sim}{\hat{Q}}),\underset{\sim}{\hat{S}},\underset{\sim}{\hat{T}}> \textit{ before } [S_c].$$

Let $Q_b \overset{def}{=} f_{QS}(S_c)$; then by the above ordering and (3),

(4) $$\phi(\underset{\sim}{\hat{Q}}) \vee (\underset{\sim}{\hat{S}},[\phi(Q_b)]).$$

We now show that $\underset{\sim}{\hat{S}} \vee \phi(\underset{\sim}{\hat{Q}})$. If we suppose the contrary; namely, that there is a particle $\underset{\sim}{R}$ such that

$$\underset{\sim}{\hat{R}} \vee (\phi(\underset{\sim}{\hat{Q}}),[S_c]) \text{ and } \underset{\sim}{R} \nparallel \underset{\sim}{S},$$

then, as above, for any $T_x \in \underset{\sim}{T}$ with $T_x < S_c$,

$$f_{TR} \circ f_{RT}(T_x) < f_{T\phi(Q)} \circ f_{\phi(Q)T}(T_x) \text{ and}$$

$$f_{TS} \circ f_{ST}(T_x) < f_{TR} \circ f_{RT}(T_x),$$

whence $$f_{TQ} \circ f_{QT}(T_x) = f_{T\phi(Q)} \circ f_{\phi(Q)T}(T_x)$$

$$> f_{TR} \circ f_{RT}(T_x) = f_{T\phi(R)} \circ f_{\phi(R)T}(T_x)$$

$$> f_{TS} \circ f_{ST}(T_x) = f_{TU} \circ f_{UT}(T_x), \text{ since } \underset{\sim}{U} \in \phi(\underset{\sim}{\hat{S}}),$$

so

$$<\underset{\sim}{U}, \phi(\underset{\sim}{\hat{R}}), \underset{\sim}{\hat{Q}}> \textit{ before } [S_c] \text{ and } \phi(\underset{\sim}{\hat{R}}) \neq \underset{\sim}{\hat{U}},$$

which contradicts (1) and shows that one of the suppositions is false. If the first supposition is false there is nothing further to prove; if the second supposition is false, we have shown that

(5) $$\underset{\sim}{\hat{S}} \vee \phi(\underset{\sim}{\hat{Q}}).$$

Consider a reflection mapping θ such that

$$\theta(\underset{\sim}{S}) \simeq \underset{\sim}{S}\ .$$

By the preceding theorem and (4),

(6) $$\theta \circ \phi(\underset{\sim}{\hat{Q}}) \vee (\underset{\sim}{\hat{S}}, [\theta \circ \phi(Q_b)]).$$

We now show that $\underset{\sim}{\hat{S}} \vee \theta \circ \phi(\underset{\sim}{\hat{Q}})$. If we suppose the contrary, namely that there is a particle $\underset{\sim}{R}'$ such that

$$\underset{\sim}{\hat{R}}' \vee (\theta \circ \phi(\underset{\sim}{\hat{Q}}), [S_c]) \text{ and } \underset{\sim}{R}' \nparallel \underset{\sim}{S},$$

then, by an argument similar to that of the preceding

paragraph, for any instant $S_x \in \underset{\sim}{S}$ and $S_x < S_c$,

$$f_{S\theta\circ\phi(Q)} \circ f_{\theta\circ\phi(Q)S}(S_x) = f_{S\phi(Q)} \circ f_{\phi(Q)S}(S_x)$$

$$> f_{SR'} \circ f_{R'S}(S_x) = f_{S\theta(R')} \circ f_{\theta(R')S}(S_x),$$

so $\hat{\underset{\sim}{S}} \not\vee \phi(\hat{\underset{\sim}{Q}})$; but this contradicts (5), whence

(7) $$\hat{\underset{\sim}{S}} \vee (\theta \circ \phi(\hat{\underset{\sim}{Q}}),[S_c]).$$

Let $S_a \overset{def}{=} f_{S\phi(Q)} \circ f_{\phi(Q)S}(S_c) = f_{S\theta\circ\phi(Q)} \circ f_{\theta\circ\phi(Q)S}(S_c)$,

$S'_a \overset{def}{=} f_{SQ} \circ f_{QS}(S_c)$ and

$T_a \overset{def}{=} f_{TQ} \circ f_{QT}(T_c) = f_{T\phi(Q)} \circ f_{\phi(Q)T}(T_c)$,

where $T_c \in \underset{\sim}{T}$ and $T_c \simeq S_c$.

By (1) and (2), $<\underset{\sim}{U},\underset{\sim}{T},\underset{\sim}{S},\underset{\sim}{Q}>$ *after* $[S_c]$, and since $\underset{\sim}{U} \in \phi(\hat{\underset{\sim}{S}})$ it follows from Theorem 19 (§4.3) that

(8) $$<\phi(\hat{\underset{\sim}{Q}}),\hat{\underset{\sim}{U}},\hat{\underset{\sim}{T}},\hat{\underset{\sim}{S}},\hat{\underset{\sim}{Q}}> \textit{ after } [S_c].$$

Since θ and ϕ are reflections, it follows that $\hat{\underset{\sim}{Q}}$ and $\theta \circ \phi(\hat{\underset{\sim}{Q}})$ are on the same side of $[S_c]$ in *col*. By (1), $\underset{\sim}{S} \nparallel \underset{\sim}{T}$, so $S'_a < T_a < S_a$, and therefore by Theorem 19 (§4.3),

$$|[S_c],[Q_b],\theta \circ \phi[Q_b]> \text{ and } \theta \circ \phi(\hat{\underset{\sim}{Q}}) \neq \hat{\underset{\sim}{Q}}.$$

Since both $\hat{\underset{\sim}{Q}}$ and $\theta \circ \phi(\hat{\underset{\sim}{Q}})$ diverge from $\underset{\sim}{S}$, Theorem 35 (§7.1) implies that there is no event at which $\hat{\underset{\sim}{Q}}$ and $\theta \circ \phi(\hat{\underset{\sim}{Q}})$ can coincide. Thus by (8) and Corollary 1 of Theorem 33 (§6.4),

$$<\phi(\hat{\underset{\sim}{Q}}),\hat{\underset{\sim}{S}},\hat{\underset{\sim}{T}},\hat{\underset{\sim}{U}},\hat{\underset{\sim}{Q}},\theta \circ \phi(\hat{\underset{\sim}{Q}})> \textit{ before } [S_c].$$

This is a contradiction of (7) since by (1), $\underset{\sim}{S} \nparallel \underset{\sim}{U}$. We conclude that the supposition (1) was false, which completes the proof for the case of divergent parallels.

Case b. Convergent Parallels.

A similar proof can be based on a figure which is a reflection of Fig. 46 in a "horizontal" line. □

THEOREM 38 (Transitivity of Parallelism)

Let $\underset{\sim}{Q},\underset{\sim}{R},\underset{\sim}{S} \in COL$. If $\underset{\sim}{Q} \parallel \underset{\sim}{R}$ and $\underset{\sim}{R} \parallel \underset{\sim}{S}$, then $\underset{\sim}{Q} \parallel \underset{\sim}{S}$.

This theorem is a consequence of Theorems 17 (§4.2), 35 (§7.1) and 37 (§7.2). It is used implicitly in many subsequent theorems.

PROOF. This proof is analogous to the proof of the corresponding theorem of absolute geometry.

Case (a) Divergent Parallels

The result is trivial unless $\underset{\sim}{Q} \nparallel \underset{\sim}{R} \nparallel \underset{\sim}{S} \nparallel \underset{\sim}{Q}$ which is assumed from now on. We define the right side of $\underset{\sim}{Q}$ to be the side which contains $\underset{\sim}{R}$. Now $\underset{\sim}{Q} \nparallel \underset{\sim}{S}$, so Theorem 35 (§7.1) implies that there is no event at which $\underset{\sim}{Q}$ and $\underset{\sim}{S}$ can coincide.

Case (a)(i) $\langle \underset{\sim}{Q}, \underset{\sim}{R}, \underset{\sim}{S} \rangle$ (see Fig. 47)

Take any instant $Q_c \in \underset{\sim}{Q}$, and any particle $\underset{\sim}{T} \in SPR[Q_c] \cap COL$ such that $\underset{\sim}{T}$ is on the right side of $\underset{\sim}{Q}$ before $[Q_c]$. Since $\underset{\sim}{Q} \vee \underset{\sim}{R}$, $\underset{\sim}{T}$ coincides with (and crosses) $\underset{\sim}{R}$ at some event $[R_b]$ where $R_b \in \underset{\sim}{R}$ and $[R_b] < [Q_c]$. Similarly, since $\underset{\sim}{R} \vee \underset{\sim}{S}$ and $\underset{\sim}{T}$ is on the right side of $\underset{\sim}{R}$ before $[R_b]$, $\underset{\sim}{T}$ coincides with $\underset{\sim}{S}$ at some event $[S_a]$ where $S_a \in \underset{\sim}{S}$ and $[S_a] < [R_b]$. Since $\underset{\sim}{Q}$ and $\underset{\sim}{S}$ coincide at no event, we conclude that $\underset{\sim}{Q} \vee \underset{\sim}{S}$.

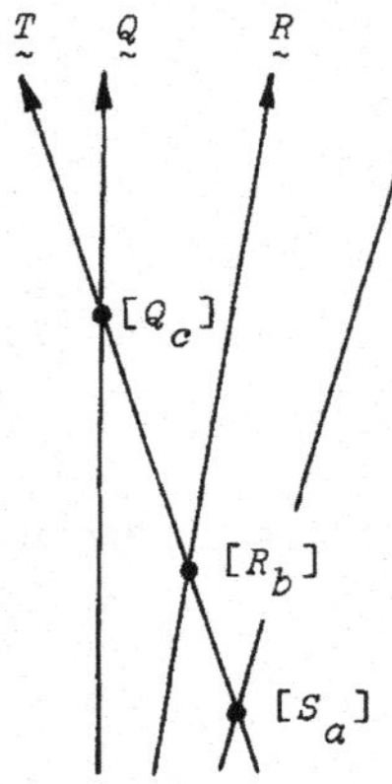

Fig. 47

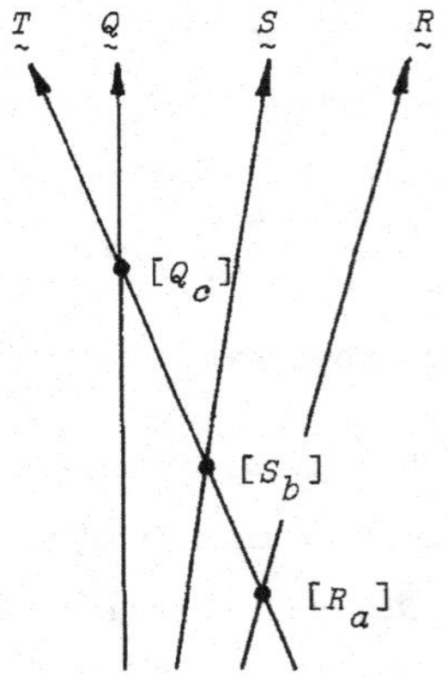

Fig. 48

Case (a)(ii) $\langle \underset{\sim}{Q}, \underset{\sim}{S}, \underset{\sim}{R} \rangle$ (See Fig. 48)

Take any instant $Q_c \in \underset{\sim}{Q}$ and any particle $\underset{\sim}{T} \in SPR[Q_c] \cap COL$ such that $\underset{\sim}{T}$ is on the right side of $\underset{\sim}{Q}$ before $[Q_c]$. Since $\underset{\sim}{Q} \vee \underset{\sim}{R}$, the particle $\underset{\sim}{T}$ coincides with $\underset{\sim}{R}$ at some event $[R_a]$ where $R_a \in \underset{\sim}{R}$ and $[R_a] < [Q_c]$. By Theorem 17 (§4.2) the particle $\underset{\sim}{T}$ coincides with $\underset{\sim}{S}$ at some event $[S_b]$ where $S_b \in \underset{\sim}{S}$ and $[R_a] < [S_b] < [Q_c]$. Since $\underset{\sim}{Q}$ and $\underset{\sim}{S}$ coincide at no event we conclude that $\underset{\sim}{Q} \vee \underset{\sim}{S}$.

Case (a)(iii) $\langle \underset{\sim}{S}, \underset{\sim}{Q}, \underset{\sim}{R} \rangle$

The previous theorem implies that $\underset{\sim}{S} \vee \underset{\sim}{R}$ and $\underset{\sim}{R} \vee \underset{\sim}{Q}$ so interchanging the symbols "Q" and "S" in case (a) (ii) , we find that $\underset{\sim}{S} \vee \underset{\sim}{Q}$. Again by the previous theorem, $\underset{\sim}{Q} \vee \underset{\sim}{S}$. This completes the proof for divergent parallels.

Case (b) Convergent Parallels

A similar proof applies with the word "before" and the symbols $<$ and $\vee$ replaced by "after" and $>$ and $\wedge$, respectively. □

§7.3 Coordinates on a Collinear Set

By considering any equivalence class of parallels in *COL*, we can attach "coordinates" to the events of *col*.

Before discussing classes of parallel particles we first consider those subclasses of parallels which can be "indexed" by dyadic numbers (recall that a *dyadic number* is a number of the form $n/2^m$, where n is any integer and m is any non-negative integer).

THEOREM 39 (Existence of Mid-Way Parallel)
Let $\underset{\sim}{Q}, \underset{\sim}{U}$ be distinct particles with $\underset{\sim}{Q} \parallel \underset{\sim}{U}$. There exists a particle $\underset{\sim}{S}$ which is mid-way between $\underset{\sim}{Q}$ and $\underset{\sim}{U}$ and parallel to both.

This theorem is a consequence of Theorems 23 (§5.3) and 36 (§7.1). It is used in the proof of Theorems 44 (§7.4) and 47 (§7.5).

PROOF. The proof is essentially the same as the proof of Theorem 23 (§5.3) except that Theorem 36 (§7.1) takes the place of Theorems 21 (§5.1), and 22(§5.1) and Corollary 2 to Theorem 22 (§5.1); and Theorem 36 (§7.1) takes the place of Theorem 5 (§2.9). □

COROLLARY. *Let $\underset{\sim}{S}^0, \underset{\sim}{S}^1$ be distinct particles with $\underset{\sim}{S}^0 \parallel \underset{\sim}{S}^1$. There is a collinear class of parallel particles*

$$\{\underset{\sim}{S}^d : d \text{ is a dyadic number}\}$$

such that, for any integers m and n,

$$\left[f_{S^{dm}S^{dn}} \circ f_{S^{dn}S^{dm}} \right]^* = \left[f_{S^{dm}S^{d(m+1)}} \circ f_{S^{d(m+1)}S^{dm}} \right]^{n-m}$$

and for any dyadic numbers a,b,c

$$a < b < c \Longleftrightarrow \langle \underset{\sim}{S}^a, \underset{\sim}{S}^b, \underset{\sim}{S}^c \rangle .$$

This corollary is a consequence of Theorem 19 (§4.3) and the Corollary to Theorem 24 (§5.3). It is used in the proof of Theorem 41 (§7.3).

PROOF. By the above theorem there is a particle $\underset{\sim}{S}^{\frac{1}{2}}$ mid-way between $\underset{\sim}{S}^0$ and $\underset{\sim}{S}^1$ and by induction there is a particle

(1) $\qquad \underset{\sim}{S}^{2^{-(m+1)}}$ mid-way between $\underset{\sim}{S}^0$ and $\underset{\sim}{S}^{2^{-m}}$.

We define the right side of $\underset{\sim}{S}^0$ to be the side which contains $\underset{\sim}{S}^1$. In the remainder of this proof we shall use the notation:

$$\rho\langle a;b\rangle \overset{def}{=} \left[f_{S^aS^b} \circ f_{S^bS^a} \right]^* ,$$

where the superscripts a and b are not necessarily numbers. As in the Corollary to Theorem 24 (§5.3), for each integer p the set of particles

$$\{\underset{\sim}{S}^{n/2^p} : n=0,\pm 1,\pm 2,\cdots\}$$

has the property:

(2) $$\rho\langle m/2^p;n/2^p\rangle = \rho^{n-m}\langle m/2^p;(m+1)/2^p\rangle .$$

Now by (1),

$$\rho^2\langle 0;1/2^{(p+1)}\rangle = \rho\langle 0;1/2^p\rangle ,$$

and so by induction,

$$\rho^{2^q}\langle 0;1/2^{(p+q)}\rangle = \rho\langle 0;1/2^p\rangle .$$

Therefore, for any integer n,

$$\rho^{n.2^q}\langle 0;1/2^{(p+q)}\rangle = \rho^n\langle 0;1/2^p\rangle ,$$

and by (2),

$$\rho\langle 0;n.2^q/2^{(p+q)}\rangle = \rho\langle 0;n/2^p\rangle .$$

Now by Theorem 19 (§4.3), for all integers n and for any non-negative integers p and q,

$$\underset{\sim}{S}^{n.2^q/2^{(p+q)}} \simeq \underset{\sim}{S}^{n/2^p} ;$$

that is, the set of particles

$$\{\underset{\sim}{S}^{n.2^q/2^{(p+q)}} : q=0,1,2,\cdots\}$$

is an equivalence class of permanently coincident particles. Equation (2) is equivalent to the equation which was to be proved; the ordering property is trivial. □

A subclass of (convergent or divergent) parallels indexed by dyadic numbers will be called a *dyadic class of parallels*. We can define a *dyadic class of instants* of the particle $\underset{\sim}{S}^0$, by taking any particular instant of $\underset{\sim}{S}^0$ and giving it the index S_0^0, and then letting

$$S_{2p}^0 \overset{def}{=} \left(f_{S^0 S^p} \circ f_{S^p S^0} \right)^* (S_0^0) ,$$

for each dyadic number p. If it is clear from the context that we are referring to a particular class of parallels and instants we shall simply call them *dyadic parallels* and *dyadic instants*, respectively.

A further consequence of the preceding c.rollary is that the dyadic subscripts (of the subset of dyadic instants of the particle $\underset{\sim}{S}^0$) are ordered in accordance with the ordering of the instants they represent; that is, for any dyadic numbers a and b,

$$a < b \Leftrightarrow S_a^0 < S_b^0 .$$

THEOREM 40. *Any dyadic class of instants of any particle is a countable dense subset of (the set of instants of) the particle.*

This theorem is a consequence of Theorems 12 (§3.5), 15 (§3.7), 19 (§4.3), 35 (§7.1), 36 (§7.1) and the Corollary to Theorem 39 (§7.3). It is used in the proof of Theorems 41 (§7.3) and 46 (§7.5).

PROOF. Let $\underset{\sim}{S}^0$ be a given particle with a given subset of dyadic instants, and let $\underset{\sim}{T} \in \hat{\underset{\sim}{S}}^0$. Define a sequence of instants

$$\left\{T_n : T_n \simeq S^0_{(2-1/2^n)} : n=0,1,\cdots;\ T_n \in \underset{\sim}{T}\right\}$$

This sequence is bounded and strictly increasing and therefore has a supremum

$$(1) \qquad T^0_\omega \simeq S^0_\omega \overset{def}{=} sup\left\{S^0_{(2-1/2^n)} : n=0,1,2,\cdots\right\}.$$

We will first show that $S^0_\omega = S^0_2$.

Let $S^1_{-1} \overset{def}{=} f^{-1}_{S^0S^1}(S^0_0)$ and let $S^1_1 \overset{def}{=} f_{S^1S^0}(S^0_0)$.

By the previous corollary,

$$(2) \qquad \left[f_{S^1 S^{(1-1/2^n)}} \circ f_{S^{(1-1/2^n)} S^1} \right]^{2^n} (S^1_{-1}) = S^1_1$$

By Theorem 36 (§7.1) there is a parallel $\underset{\sim}{Q}$ (convergent or divergent as the case may be) such that

(3) $$f_{S^0 Q} \circ f_{Q S^0}(S^0_0) = S^0_\omega$$

and since $S^0_\omega = \sup_n\left\{S^0_{2-1/2^n} \leqslant S^0_2\right\}$, it follows that for all positive integers n,

$$<\underset{\sim}{S}^0, \underset{\sim}{S}^{1-1/2^n}, \underset{\sim}{Q}, \underset{\sim}{S}^1>$$

So by equation (2) and Theorem 19 (§4.3), it follows that for all positive integers n,

$$\left(f_{S^1 Q} \circ f_{Q S^1}\right)^{2^n}(S^1_{-1}) \leqslant S^1_1 ,$$

whence

$$\sup_n\left(f_{S^1 Q} \circ f_{Q S^1}\right)^{2^n}(S^1_{-1}) \leqslant S^1_{-1} ,$$

and so by Theorem 12 (§3.5) the particles $\underset{\sim}{Q}$ and $\underset{\sim}{S}^1$ coincide at the event

$$\left[\sup_n\left\{\left(f_{S^1 Q} \circ f_{Q S^1}\right)^{2^n}(S^1_{-1})\right\}\right]$$

Therefore $\underset{\sim}{Q} \simeq \underset{\sim}{S}^1$ and so by (3),

(4) $$S^0_\omega = S^0_2 .$$

The remaining part of this proof is based on the proof of a theorem of Walker [1948, Theorem 13.1, P330]. Given any two instants $S^0_x, S^0_z \in \underset{\sim}{S}^0$ with $S^0_x < S^0_z$, we will find a dyadic instant S^0_y such that $S^0_x < S^0_y < S^0_z$. Theorem 15 (§3.7) and

Theorem 36 (§7.1) imply that there is a parallel $\underset{\sim}{U}$ (divergent or convergent as the case may be) to the right of $\underset{\sim}{S}^0$ such that

(5) $$S^0_x < u^{-1}(S^0_z) < S^0_z ,$$

where $u \overset{def}{=} f_{S^0U} \circ f_{US^0}$.

By Theorem 35 (§7.1) $\underset{\sim}{U}$ can not coincide with $\underset{\sim}{S}^0$ at $[S^0_2]$ and by equation (1) there is some positive integer m such that

(6) $$S^0_{(2-2^{-m+2})} > u^{-1}(S^0_2) .$$

We define the set of particles

$$\{\underset{\sim}{R}^{(n)}: \underset{\sim}{R}^{(n)} \simeq \underset{\sim}{S}^{2^{-n}}; n=1,2,3,\cdots\}$$

and so

(7) $$r_n^{-1}(S^0_2) = S^0_{(2-2^{-n+1})} > S^0_{(2-2^{-n+2})} ,$$

where $r_n \overset{def}{=} f_{S^0R^{(n)}} \circ f_{R^{(n)}S^0}$.

By (6) and (7),

$$r_m^{-1}(S^0_2) > u^{-1}(S^0_2)$$

and since parallels can not cross (by Theorem 35 (§7.1))

$$r_m^{-1} > u^{-1} .$$

Thus from (5),

(8) $$S_x^0 < r_m^{-1}(S_z^0) < S_z^0 \ .$$

Since $\underset{\sim}{R}^{(m)}$ does not coincide with $\underset{\sim}{S}^0$ at any event, the increasing and decreasing sequences

$$\left\{r_m^n(S_2^0):\ n=0,1,2,\cdot\cdot\right\} \ \ and \ \left\{r_m^{-n}(S_2^0):\ n=0,1,2,\cdot\cdot\right\}$$

are unbounded, so (the set of instants of) $\underset{\sim}{S}^0$ is covered by the set of semi-closed intervals

$$\left\{\left(r_m^n(S_2^0), r_m^{n+1}(S_2^0)\right]:\ n=0,\pm 1,\pm 2,\ldots\right\}$$

Therefore there is some integer p such that

(9) $$r_m^p(S_2^0) < S_z^0 \leqslant r_m^{p+1}(S_2^0) \ .$$

Now from (8) and (9),

$$S_x^0 < r_m^{-1}(S_z^0) \leqslant r_m^{-1} \circ r_m^{p+1}(S_2^0) = r_m^p(S_2^0) < S_z^0$$

and by definition, $r_m^p(S_2^0)$ is a dyadic instant. □

COROLLARY. *The set of instants of each particle is order-isomorphic to the set of real numbers.*

This corollary is a consequence of Theorems 11 (§3.4) and 25 (§5.4). It is used implicitly in many of the following theorems.

PROOF. By Theorem 11 (§3.4) particles do not have first or last instants. By the above theorem each particle has a countable dense subset of instants. By Theorem 25 (§5.4) the set of instants of each particle has no gaps (defined in §3.1). Sierpinski [1965, $\overline{\text{XI}}$, §10, Theorem 1] has shown that these conditions are necessary and sufficient for a linearly-ordered set to be order-isomorphic to the reals. □

THEOREM 41 (Indexed Class of Parallels)
A class of parallels and the instants belonging to them can be indexed by the real numbers such that, for any real numbers a,b,c

(i) $$f^{+}_{cb}(S^{b}_{a}) = S^{c}_{a-b+c} \quad \text{and} \quad f^{-}_{cb}(S^{b}_{a}) = S^{c}_{a+b-c} \, ,$$

whence

(ii) $$\left(f_{bc} \circ f_{cb}\right)^{*}(S^{b}_{a}) = S^{b}_{a-2b+2c} \, ,$$

where we have introduced the notation: $f_{ab} \overset{\text{def}}{=} f_{S^a S^b}$.

(see Fig. 49).

Furthermore, for any real numbers a and b, there is a parallel $\underset{\sim}{S}^{b}$ *and an instant* $S^{b}_{a} \in \underset{\sim}{S}^{b}$. *A class of parallels, indexed in this way, is called an indexed class of parallels. The subscript indices of any parallel are said to constitute a divergent or convergent time scale, according as to whether the class of parallels is a divergent or convergent class.*

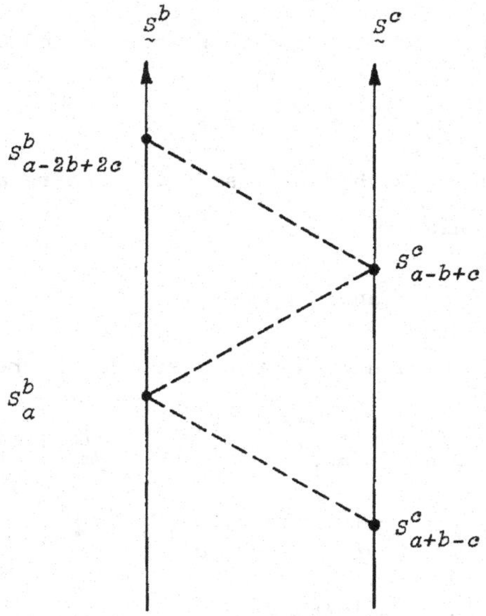

Fig. 49

This theorem is a consequence of Theorems 9 (§3.2), 18 (§4.3), 19 (§4.3), 36 (§7.1), the Corollary to Theorem 39 (§7.3) and Theorem 40 (§7.3). It is used in the proof of Theorems 45 (§7.4), 46 (§7.5), 47 (§7.5), 48 (§7.5), 49 (§7.5) and the Corollary to Theorem 56 (§8.4).

PROOF. We have already indexed the dyadic instants of the particle $\underset{\sim}{S}^0$ by letting

$$(1) \qquad S^0_{2\gamma} \overset{def}{=} \left[f_{0\gamma} \circ f_{\gamma 0} \right]^* (S^0_0) .$$

We will first show that, for any dyadic numbers α and β,

$$(2) \qquad \left(f_{0\beta} \circ f_{\beta 0}\right)^{*}(S^{0}_{2\alpha}) = S^{0}_{2\alpha+2\beta} \, .$$

Let d be a dyadic number (with $d \leqslant 1$) such that there are integers m and n for which

$$\alpha = dm \quad \text{and} \quad \beta = dn \, .$$

Thus, by the definition (1) and the Corollary to Theorem 39, (§7.3),

$$\begin{aligned}
\left(f_{0\beta} \circ f_{\beta 0}\right)^{*}(S^{0}_{2\alpha}) &= \left(f_{0\,dn} \circ f_{dn\,0}\right)^{*} \circ \left(f_{0\,dm} \circ f_{dm\,0}\right)^{*}(S^{0}_{0}) \\
&= \left(f_{0d} \circ f_{d0}\right)^{n} \circ \left(f_{0d} \circ f_{d0}\right)^{m}(S^{0}_{0}) \\
&= \left(f_{0d} \circ f_{d0}\right)^{m+n}(S^{0}_{0}) \\
&= \left(f_{0\,d(m+n)} \circ f_{d(m+n)\,0}\right)^{*}(S^{0}_{0}) \\
&= S^{0}_{2d(m+n)} \\
&= S^{0}_{2\alpha+2\beta} \, ,
\end{aligned}$$

which establishes equation (2).

We now extend this result to apply to all instants of all parallels. By the previous theorem, any instant of $\underset{\sim}{S}^{0}$ specifies a (Dedekind) cut in the dense subset of dyadic instants of $\underset{\sim}{S}^{0}$.

We index each instant of $\underset{\sim}{S}^0$ by the real number so obtained; also by the Monotonic Sequence Theorem (Theorem 9, §3.1), for each real number there is a corresponding instant of $\underset{\sim}{S}^0$; thus for each real number a and for any dyadic numbers α and $\overline{\alpha}$,

(3) $$\alpha < a < \overline{\alpha} \iff S^0_{\alpha} < S^0_{a} < S^0_{\overline{\alpha}} \,.$$

Furthermore, any class of parallels is a permanently ordered set, and by Theorem 19 (§4.3), between any two distinct parallels there is some dyadic parallel. In much the same way as before, any parallel specifies a (Dedekind) cut (see §3.1) in the dense subset of dyadic parallels. We index each parallel by the real number so obtained; also by Theorem 36 (§7.1) and Theorem 19 (§4.3), for each real number b there is a parallel $\underset{\sim}{S}^b$ such that

$$\left(f_{0b} \circ f_{b0}\right)^*(S^0_0) = S^0_{2b} \;;$$

thus, as above, for each real number b, and for all dyadic numbers β and $\overline{\beta}$,

(4) $$\beta < b < \overline{\beta} \iff \left(f_{0\beta} \circ f_{\beta 0}\right)^* < \left(f_{0b} \circ f_{b0}\right)^* < \left(f_{0\overline{\beta}} \circ f_{\overline{\beta}0}\right)^* \,.$$

Thus, by equations (2) and (3), for any given real number a and for all dyadic numbers α and $\overline{\alpha}$ such that $\alpha < a < \overline{\alpha}$,

$$S^0_{2\alpha+2\beta} < \left(f_{0\beta} \quad f_{\beta 0}\right)^*(S^0_{2a}) < S^0_{2\overline{\alpha}+2\beta} \,,$$

whence

$$\left(f_{0\beta} \circ f_{\beta 0}\right)^{*}(S^{0}_{2a}) = S^{0}_{2a+2\beta} .$$

Similarly, by the above equation and (4), for any real number b and for all dyadic numbers β and $\overline{\beta}$ such that $\beta < b < \overline{\beta}$,

$$S^{0}_{2a+2\beta} < \left(f_{0b} \circ f_{b0}\right)^{*}(S^{0}_{2a}) < S^{0}_{2a+2\overline{\beta}} ,$$

whence

(5) $$\left(f_{0b} \circ f_{b0}\right)^{*}(S^{0}_{2a}) = S^{0}_{2a+2b} .$$

We can now index the instants of each particle by defining, for all real numbers a and c,

(6) $$S^{c}_{a} \overset{def}{=} f^{+}_{c0}(S^{0}_{a-c}) .$$

Now with the instants of the parallels indexed in this way, equation (5) and Theorem 18 (§4.3) imply that

(7) $$S^{0}_{a+c} = f^{-}_{0c}(S^{c}_{a}) .$$

Furthermore, by Theorem 18 (§4.3)

$$f^{+}_{cb} = f^{+}_{c0} \circ \left(f^{+}_{b0}\right)^{-1} \quad \text{and} \quad f^{-}_{cb} = f^{-}_{c0} \circ \left(f^{-}_{b0}\right)^{-1} ,$$

so by (6) and (7),

(8) $$f^{+}_{cb}(S^{b}_{a}) = S^{c}_{a-b+c} \quad \text{and} \quad f^{-}_{cb}(S^{b}_{a}) = S^{c}_{a+b-c} ,$$

whence, again by Theorem 18 (§4.3),

$$(9) \qquad \left(\underset{bc}{f} \circ \underset{cb}{f}\right)^{*}(S^b_a) = S^b_{a-2b+2c} \,.$$

This, together with equations (8), completes the proof. □

COROLLARY. *Let* $\underset{\sim}{Q} \in COL$ *and let* $[U_a] \in col$. *If the two types of parallel to* $\underset{\sim}{Q}$ *through* $[U_a]$ *coincide, then there is only one distinct parallel to* $\underset{\sim}{Q}$ *through each event of col.*

This corollary is a consequence of Theorems 19 (§4.3) and 36 (§7.1). It is used in the proof of Theorem 46 (§7.5).

PROOF. (see Fig. 50).
Let $\{\underset{\sim}{R}^{\alpha}: \alpha \text{ real}, \underset{\sim}{R}^{\alpha} \in COL\}$ *and* $\{\underset{\sim}{S}^{\beta}: \beta \text{ real}, \underset{\sim}{S}^{\beta} \in COL\}$ *be* classes of divergent and convergent parallels such that $\underset{\sim}{Q} \approx \underset{\sim}{R}^{0} \approx \underset{\sim}{S}^{0}$. Then through $[U_a]$ there are divergent and convergent parallels $\underset{\sim}{R}^{k}$ and $\underset{\sim}{S}^{l}$, respectively, where k and l are real numbers, such that $\underset{\sim}{R}^{k} \approx \underset{\sim}{S}^{l}$; so by considering record functions and applying Theorem 19 (§4.3), we find that, for each integer n,

$$\underset{\sim}{R}^{nk} \approx \underset{\sim}{S}^{nl} \,.$$

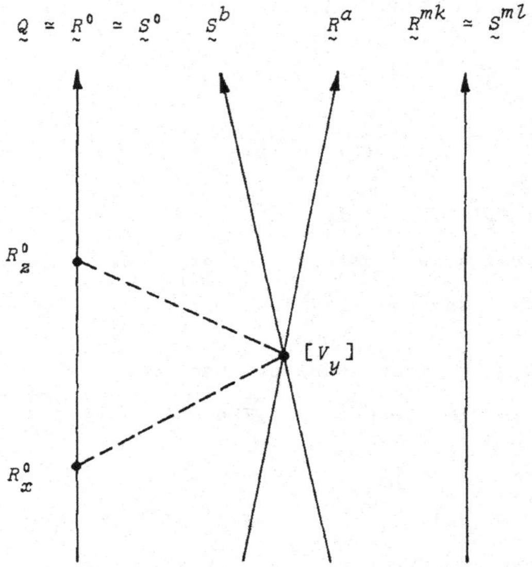

Fig. 50 In this diagram, $0 < z-x < m$.

Now take any event $[V_y] \in col$ and let

$$R_x^0 \overset{def}{=} \underset{R^0V}{f^+}(V_y) \quad \text{and} \quad R_z^0 \overset{def}{=} \underset{R^0V}{f^-}(V_y) \ .$$

If $[V_y]$ is to the left of $\underset{\sim}{Q}$, $z < x$ and we take an integer m such that $m < z-x < 0$; if $[V_y]$ is to the right of $\underset{\sim}{Q}$, $z > x$ and we take an integer m such that $0 < z-x < m$ (see Fig. 50).

By Theorem 36 (§7.1), there are parallels $\underset{\sim}{R}^a, \underset{\sim}{S}^b$ such that

(1) $$\underset{\sim}{R}^a \vee (\underset{\sim}{R}^0, [V_y]) \text{ and } \underset{\sim}{S}^b \wedge (\underset{\sim}{S}^0, [V_y]) \therefore$$

Since the relations of parallelism are equivalence relations,

(2) $$\underset{\sim}{R}^a \vee \underset{\sim}{R}^0 \text{ and } \underset{\sim}{R}^0 \vee \underset{\sim}{R}^{mk} \Rightarrow \underset{\sim}{R}^a \vee \underset{\sim}{R}^{mk} \text{ and}$$

$$\underset{\sim}{S}^b \wedge \underset{\sim}{S}^0 \text{ and } \underset{\sim}{S}^0 \wedge \underset{\sim}{S}^{ml} \text{ and } \underset{\sim}{R}^{mk} \simeq \underset{\sim}{S}^{ml} \Longrightarrow \underset{\sim}{S}^b \wedge \underset{\sim}{R}^{mk}$$

Now (1) and (2) imply that

$$\langle \underset{\sim}{R}^0 \simeq \underset{\sim}{S}^0, \underset{\sim}{S}^b, \underset{\sim}{R}^a \rangle \text{ after } [V_y] \text{ and } \langle \underset{\sim}{R}^a, \underset{\sim}{S}^b, \underset{\sim}{R}^{mk} \rangle \text{ after } [V_y]$$

and hence $\underset{\sim}{R}^a \simeq \underset{\sim}{S}^b$. That is, there is only one particle which is parallel to $\underset{\sim}{Q}$, in both the divergent and convergent senses, which coincides with the event $[V_y]$. □

This corollary and its proof are analogous to a proposition of absolute geometry. The corollary is used in the proof of the "Euclidean" Parallel Theorem (Theorem 46 (§7.5))

The next theorem shows that for any given particle, all time scales of the same type are determined to within an arbitrary strictly increasing linear transformation. In the axiomatic system of Szekeres, this property was regarded as an axiom and was called the "Axiom of Standard Time" [Szekeres, 1968, Axiom A.8, P142]. In the present system, the property is a consequence of the Axiom of Isotropy of Sprays (Axiom VII, §2.9).

THEOREM 42 (Affine Invariance of Divergent and Convergent Time Scales)

Let COL (1) and COL (2) be collinear sets of particles, and let $\{\underset{\sim}{S}^{\alpha}\colon \alpha \text{ real}, \underset{\sim}{S}^{\alpha} \in COL\ (1)\}$ *and* $\{\underset{\sim}{U}^{\alpha}\colon \alpha \text{ real}, \underset{\sim}{U}^{\alpha} \in COL\ (2)\}$ *be classes of parallels of the same type indexed by the reals.*

(i) *If* $\underset{\sim}{S}^{a} \simeq \underset{\sim}{U}^{b}$, *then there are real constants c,d,k such that for all real x,*

$$S^{a}_{c+kx} \simeq U^{b}_{d+x} \; .$$

(ii) *Also, if COL (1) = COL (2), then for all real x and y,*

$$S^{a\pm ky}_{c+kx} \simeq U^{b+y}_{d+x} \; ,$$

the upper or lower sign being chosen according as to whether the left side of $\underset{\sim}{S}^{a}$ *is the left, or right, side of* $\underset{\sim}{U}^{b}$.

This theorem is a consequence of Axiom VII (§2.9) and Theorems 9 (§3.2), 24 (§5.3) and 35 (§7.1). It is used in the proof of Theorems 47 (§7.5), 48 (§7.5) and 49 (§7.5).

PROOF. (i) If $\underset{\sim}{S}^{a} \simeq \underset{\sim}{U}^{b}$, we consider any instant $S^{a}_{\alpha} \in \underset{\sim}{S}^{a}$; then there is some instant $U^{b}_{\beta} \in \underset{\sim}{U}^{b}$ such that

(1) $$S^{a}_{\alpha} \simeq U^{b}_{\beta} \; .$$

For any parallel $\underset{\sim}{S}^{c}$ such that $\underset{\sim}{S}^{c} \nparallel \underset{\sim}{S}^{a}$, Theorem 36 (§7.1) shows that there is a parallel $\underset{\sim}{U}^{d} \nparallel \underset{\sim}{U}^{b}$ such that

$$f_{S^{a}S^{c}} \circ f_{S^{c}S^{a}}(S^{a}_{\alpha}) \simeq f_{U^{b}U^{d}} \circ f_{U^{d}U^{b}}(U^{b}_{\beta}) \; .$$

In §3.6 we saw that signal functions could specify mappings between the sets of events coincident with particles, as well as between the particles themselves. With this interpretation,

(2) $$f_{S^a S^c} \circ f_{S^c S^a} = f_{U^b U^d} \circ f_{U^d U^b} ,$$

the proof being analogous to the proof of Theorem 36 (part (ii)) (§7.1) with reflection mappings being replaced by the more general isotropy mappings.

By (1) and (2) we see that, for any integer n,

(3) $$\left[S^a_{\alpha+2n(c-a)}\right] = \left[U^b_{\beta+2n(d-b)}\right] .$$

By considering parallels mid-way between $\underset{\sim}{S}^a$ and $\underset{\sim}{S}^c$, and mid-way between $\underset{\sim}{U}^b$ and $\underset{\sim}{U}^d$, it follows that

$$\left(f_{S^a S^{(a+c)/2}} \circ f_{S^{(a+c)/2} S^a}\right)^2 = \left(f_{U^b U^{(b+d)/2}} \circ f_{U^{(b+d)/2} U^b}\right)^2$$

Since these record functions are continuous functions of a real variable, the intermediate value theorem (see, for example, Fulks [1961, Theorem 3.3a]) applies and so there are instants $S^a_x \in \underset{\sim}{S}^a$ and $U^b_y \in \underset{\sim}{U}^b$ such that

$$[S^a_x] = [U^b_y]$$

and

$$f_{S^a S^{(a+c)/2}} \circ f_{S^{(a+c)/2} S^a} [S^a_x] = f_{U^b U^{(b+c)/2}} \circ f_{U^{(b+c)/2} U^b} [U^b_y] .$$

Then, as with (2) above, for any integer n it follows that

$$\left[S^a_{\alpha+2n(c-a)/2}\right] = \left[U^b_{\beta+2n(b-d)/2}\right] .$$

Proceding by induction we see that, for all positive integers m and for all integers n,

$$\left[S^a_{\alpha+2n(c-a)/2^m}\right] = \left[U^b_{\beta+2n(d-b)/2^m}\right] .$$

Thus, letting $k \overset{def}{=} (c-a)/(d-b)$ and noting that the indices of $\underset{\sim}{U}^b$ are mapped onto the indices of $\underset{\sim}{S}^a$ by some strictly monotonic increasing function we see that, for all real x,

(4) $$\left[S^a_{\alpha+kx}\right] = \left[U^b_{\beta+x}\right] .$$

(ii) If $COL\ (1) = COL\ (2)$, then according as to whether the left side of $\underset{\sim}{S}^a$ is the left, or right, side of $\underset{\sim}{U}^b$, Theorem 19 (§4.3) implies that, for all real y,

$$\underset{\sim}{S}^{a\underline{+}ky} \simeq \underset{\sim}{U}^{b+y}$$

and so by (4) and the relations given by the previous theorem it follows that, for all real x and y,

$$\left[S^{a\pm ky}_{\alpha+kx}\right] = \left[U^{b+y}_{\beta+x}\right] .$$

§7.4 Isomorphisms of a Collinear Set of Particles

We will now apply a reflection operation to any collinear set of particles. By composing these reflection operations we can obtain mappings which are like space translations and pseudo-rotations. Finally by composing four of these mappings, we can generate time translation mappings.

We first show that any *COL* can be reflected in any of its members and that this automorphism preserves both relations of parallelism.

THEOREM 43 (Invariance of Parallelism)

Let COL be a collinear set of particles containing the distinct particles $\underset{\sim}{R}$ *and* $\underset{\sim}{V}$*, together with a particle* $\underset{\sim}{T}$ *mid-way between* $\underset{\sim}{R}$ *and* $\underset{\sim}{V}$ *such that either:*

(i) *there are instants* $R_c \in \underset{\sim}{R}$ *and* $V_c \in \underset{\sim}{V}$ *such that* $R_c \simeq V_c$*, or*

(ii) $\underset{\sim}{R} \vee \underset{\sim}{V}$ *, or*

(iii) $\underset{\sim}{R} \wedge \underset{\sim}{V}$.

For any particle $\underset{\sim}{S} \in COL$ *such that* $\underset{\sim}{S} \parallel \underset{\sim}{R}$*, there is an observer* $\underset{\sim}{S}_T \subset COL$ *and* $\hat{\underset{\sim}{S}}_T \parallel \hat{\underset{\sim}{R}}_T$.

This theorem is a consequence of Theorems 18 (§4.3), 19 (§4.3) and 36 (§7.1). It is used in the proof of Theorem 44 (§7.4).

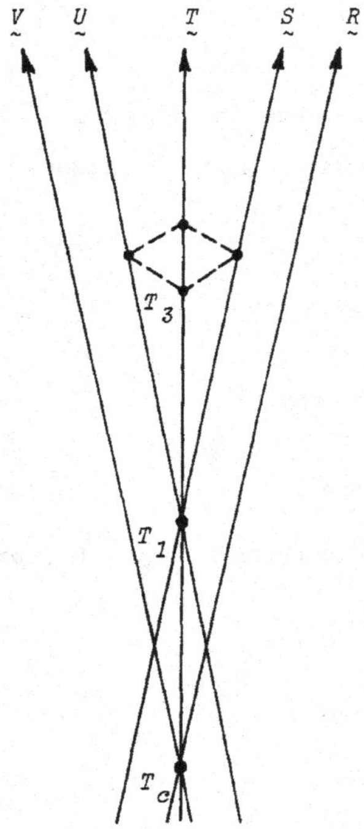

Fig. 51

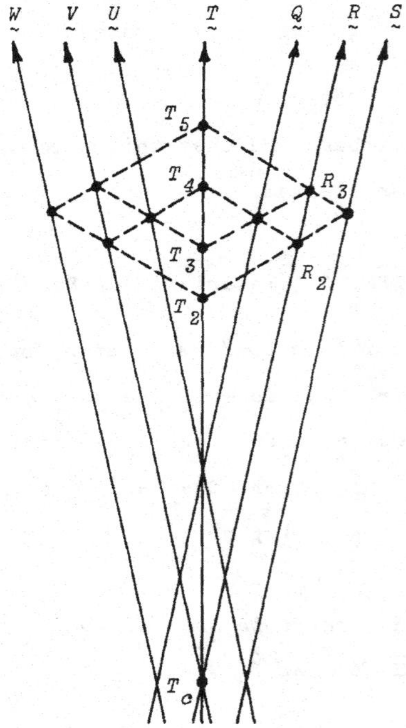

Fig. 52

PROOF. There are six cases to be considered, since $\underset{\sim}{R}$ and $\underset{\sim}{V}$ could be either: (i) members of a SPRAY, or (ii) divergent parallels, or (iii) convergent parallels; and $\underset{\sim}{S}$ and $\underset{\sim}{R}$ could be either: (A) convergent parallels or (B) divergent parallels.

Case (i). $\underset{\sim}{R}$ and $\underset{\sim}{V}$ members of a SPRAY

We define the sides of $\underset{\sim}{T}$ such that $\underset{\sim}{R}$ is on the right side of $\underset{\sim}{T}$ after $[R_c]$.

Case (i)(A). Convergent Parallels

Firstly we consider those convergent parallels on the left side of $\underset{\sim}{R}$, since these parallels necessarily cross $\underset{\sim}{T}$ at some event and this simplifies the proof.

Case (i)(A)(a). $\underset{\sim}{S}$ on the left of $\underset{\sim}{R}$ (see Fig. 51).

There is some instant $T_1 \in \underset{\sim}{T}$ with $T_1 > R_c$ such that $\underset{\sim}{S}$ coincides with $\underset{\sim}{T}$ at $[T_1]$; that is, $\underset{\sim}{S} \wedge (\underset{\sim}{R},[T_1])$. By Theorem 36 (§7.1) there is a particle $\underset{\sim}{U} \wedge (\underset{\sim}{V},[T_1])$. By Theorem 19 (§4.3) and the definition of parallelism (§7.1) it follows that

$$(1)\quad \left(f_{TS} \circ f_{ST}\right)^* < \left(f_{TR} \circ f_{RT}\right)^* = \left\{\left(f_{TV} \circ f_{VT}\right)^*\right\}^{-1} > \left\{\left(f_{TU} \circ f_{UT}\right)^*\right\}^{-1} .$$

Since $\underset{\sim}{S} \wedge (\underset{\sim}{R},[T_1])$ it follows that for all instants $T_x \in \underset{\sim}{T}$,

$$(2)\qquad f_{TS} \circ f_{ST}(T_x) \geq f_{TU_T} \circ f_{U_T T}(T_x) = f_{TU} \circ f_{UT}(T_x) .$$

Similarly, $\underset{\sim}{U} \wedge (\underset{\sim}{V},[T_1])$, from which we obtain the opposite inequality; therefore

$$(3)\qquad f_{TS} \circ f_{ST} = f_{TU} \circ f_{UT} .$$

That is,

(4) $$\hat{\underset{\sim}{S}} = \hat{\underset{\sim}{U}}_T \text{ and } \hat{\underset{\sim}{U}} = \hat{\underset{\sim}{S}}_T ,$$

and by definition $\underset{\sim}{S} \wedge \underset{\sim}{R}$ and $\underset{\sim}{U} \wedge \underset{\sim}{V}$, which completes the proof of Case (i)(A)(a).

Case (i)(A)(b). $\underset{\sim}{S}$ on the right side of $\underset{\sim}{R}$ (see Fig. 52). By Theorem 36 (§7.1), there is a particle $\underset{\sim}{Q}$ which is a reflection of $\underset{\sim}{S}$ in $\underset{\sim}{R}$, so then $\underset{\sim}{Q}$ is on the left of $\underset{\sim}{R}$ as in Case (i)(A)(a). Take any instant $R_2 \in \underset{\sim}{R}$ (note that R_2 is completely arbitrary even though in Fig. 52 it appears that $R_2 > R_c$). We now define:

(5) $$R_3 \overset{def}{=} \left(f_{RS} \circ f_{SR}\right)^*(R_2) = f^-_{RS} \circ f^+_{SR}(R_2) ,$$

(6) $$T_2 \overset{def}{=} f^+_{TR}(R_2) ,$$

(7) $$T_3 \overset{def}{=} f^+_{TR}(R_3) ,$$

(8) $$T_4 \overset{def}{=} f^-_{TR}(R_2) \text{ and}$$

(9) $$T_5 \overset{def}{=} f^-_{TR}(R_3) .$$

In the remainder of this proof we shall repeatedly use the results of Theorem 18 (§4.3). By equations (6) and (8), and then (7) and (9), we have

(10) $$\left(f_{TR} \circ f_{RT}\right)^*(T_2) = f^-_{TR} \circ f^+_{RT}(T_2) = T_4 \text{ and}$$

(11) $$\left(f_{TR} \circ f_{RT}\right)^*(T_3) = T_5 .$$

Also

$$(12)\quad \left(f_{TQ} \circ f_{QT}\right)^{*}(T_3) = f^{-}_{TQ} \circ f^{+}_{QT}(T_3)$$

$$= f^{-}_{TR} \circ f^{-}_{RQ} \circ f^{+}_{QR} \circ f^{+}_{RT}(T_3)$$

$$= f^{-}_{TR} \circ \left(f_{RQ} \circ f_{QR}\right)^{*} \circ f^{+}_{RT}(T_3)$$

$$= f^{-}_{TR} \circ \left\{\left(f_{RS} \circ f_{SR}\right)^{*}\right\}^{-1} \circ f^{+}_{RT}(T_3), \text{ since } \underset{\sim}{Q} \in \hat{\underset{\sim}{S}}_R,$$

$$= f^{-}_{TR} \circ \left\{\left(f_{RS} \circ f_{SR}\right)^{*}\right\}^{-1}(R_3)\ , \text{ by equation (7)},$$

$$= f^{-}_{TR}(R_2)\ , \text{ by equation (5)}\ ,$$

$$= T_4\ , \text{ by equation (8)}.$$

Also

$$\left(f_{TS} \circ f_{ST}\right)^{*}(T_2) = f^{-}_{TS} \circ f^{+}_{ST}(T_2)$$

$$= f^{-}_{TR} \circ f^{-}_{RS} \circ f^{+}_{SR} \circ f^{+}_{RT}(T_2)$$

$$= f^{-}_{TR} \circ \left(f_{RS} \circ f_{SR}\right)^{*} \circ f^{+}_{RT}(T_2)$$

$$= f^{-}_{TR} \circ \left(f_{RS} \circ f_{SR}\right)^{*}(R_2)\ , \text{ by equation (6)},$$

$$= f^{-}_{TR}(R_3)\ , \text{ by equation (5)},$$

$$= T_5\ , \text{ by equation (9)},$$

$$= \left(f_{TR} \circ f_{RT}\right)^*(T_3)\ , \text{ by equation (11),}$$

$$= \left(f_{TR} \circ f_{RT}\right)^* \circ \left\{\left(f_{TQ} \circ f_{QT}\right)^*\right\}^{-1}(T_4)\ , \quad \text{by equation (12),}$$

$$= \left(f_{TR} \circ f_{RT}\right)^* \circ \left\{\left(f_{TQ} \circ f_{QT}\right)^*\right\}^{-1} \circ \left(f_{TR} \circ f_{RT}\right)^* \quad (T_2)\ ,$$

by equation (10), and since T_2 was arbitrary,

$$(13) \quad \left(f_{TS} \circ f_{ST}\right)^* = \left(f_{TR} \circ f_{RT}\right)^* \circ \left\{\left(f_{TQ} \circ f_{QT}\right)^*\right\}^{-1} \circ \left(f_{TR} \circ f_{RT}\right)^* \quad .$$

By Theorem 36 (§7.1) there are particles $\underset{\sim}{U}$ and $\underset{\sim}{W}$ such that

$$\underset{\sim}{U} \in \hat{\underset{\sim}{Q}}_T \quad \text{and} \quad \underset{\sim}{W} \in \hat{\underset{\sim}{U}}_V\ .$$

By the previous case $\underset{\sim}{U} \wedge \underset{\sim}{V}$ and so $\underset{\sim}{W} \wedge \underset{\sim}{V}$. Then by a similar procedure, but with $f^-, f^+, \{(f \circ f)^*\}^{-1}$ taking the place of $f^+, f^-, \{(f \circ f)^*\}$ respectively, it follows that

$$(14) \quad \left\{\left(f_{TW} \circ f_{WT}\right)^*\right\}^{-1} = \left\{\left(f_{TV} \circ f_{VT}\right)^*\right\}^{-1} \circ \left(f_{TU} \circ f_{UT}\right)^* \circ \left\{\left(f_{TV} \circ f_{VT}\right)^*\right\}^{-1}$$

But $\underset{\sim}{V} \in \hat{\underset{\sim}{R}}_T$ and $\underset{\sim}{U} \in \hat{\underset{\sim}{Q}}_T$, so equations (13) and (14) imply that

$$\left(f_{TS} \circ f_{ST}\right)^* = \left\{\left(f_{TW} \circ f_{WT}\right)^*\right\}^{-1}\ .$$

That is,

$$(15) \qquad \hat{\underset{\sim}{S}} = \hat{\underset{\sim}{W}}_T \quad \text{and} \quad \hat{\underset{\sim}{W}} = \hat{\underset{\sim}{S}}_T\ ,$$

and by definition $\underset{\sim}{S} \wedge \underset{\sim}{R}$ and $\underset{\sim}{W} \wedge \underset{\sim}{V}$, which completes the proof of Case (i)(A)(b), and hence the proof of Case (i)(A).

Case (i)(B). Divergent Parallels

First we consider those divergent parallels on the right side of $\underset{\sim}{R}$, since these parallels necessarily cross $\underset{\sim}{T}$ at some event. The proof is analogous to the proof of Case (i)(A) with the symbols $\underset{\sim}{W}, \underset{\sim}{U}, \underset{\sim}{Q}, \underset{\sim}{S}$ being replaced by the symbols $\underset{\sim}{U}, \underset{\sim}{W}, \underset{\sim}{S}, \underset{\sim}{Q}$ respectively, with the inequalities in the corresponding version of equation (1) being reversed, and with equations (13) and (14) being written so that $\left(f_{TS} \circ f_{ST}\right)^*$ and $\left\{\left(f_{TW} \circ f_{WT}\right)^*\right\}^{-1}$ are displayed explicitly.

Case (ii). $\underset{\sim}{R} \vee \underset{\sim}{V}$.

We define the sides of $\underset{\sim}{T}$ such that $\underset{\sim}{R}$ is on the right side of $\underset{\sim}{T}$. If there is only one class of parallels (that is, if divergent parallels are also convergent parallels) then all particles under consideration are members of the same equivalence class of parallels and there is nothing further to prove. If there are two classes of parallels, then a proof is only required for the case of convergent parallels, since the relation of divergent parallelism is transitive and so the case of divergent parallels is trivial. The proof for the case of convergent parallels is similar to the proof of case (i)(A), the only differences being that both references to R_c are omitted.

Case (iii). $\underset{\sim}{R} \wedge \underset{\sim}{V}$

The proof for this case is similar to the proof for case (ii) except that the words "divergent" and "convergent" should be interchanged, and all signal functions should be replaced as follows:

$$f^+_{QR} \to \left(f^-_{RQ}\right)^{-1}, \; f^-_{QR} \to \left(f^+_{RQ}\right)^{-1}, \; \left(f_{QR} \circ f_{RQ}\right)^* \to \left\{\left(f_{QR} \circ f_{RQ}\right)^*\right\}^{-1},$$

and so on. A simpler way to imagine this is to consider that case (iii) is similar to case (ii) with "the direction of time reversed". □

COROLLARY. *Let COL be a collinear set of particles containing* $\underset{\sim}{T}$ *and let* $\{\underset{\sim}{Q}^\alpha : \alpha \text{ real}, \underset{\sim}{Q}^\alpha \in COL\}$ *be a class of parallel particles. Then there is a class of parallel observers.*

$$\{\Psi(\underset{\sim}{\hat{Q}}^\alpha) : \alpha \text{ real}, \Psi(\underset{\sim}{\hat{Q}}^\alpha) \subset COL\},$$

where Ψ *denotes a reflection in* $\underset{\sim}{T}$. *(Also for each particle* $\underset{\sim}{S} \in COL$, *there is an observer* $\hat{S}_{\underset{\sim}{T}} \subset COL$.*)*

This corollary is a consequence of Theorem 36 (§7.1) and is used in the proof of Theorem 44 (§7.4).

PROOF. Take any instant $T_x \in \underset{\sim}{T}$. By Theorem 36 (§7.1) there is a particle $\underset{\sim}{R} \in COL$ such that $\underset{\sim}{R}$ coincides with $\underset{\sim}{T}$ at $[T_x]$ and $\underset{\sim}{R} \parallel \underset{\sim}{Q}^0$ and so, for all real α, $\underset{\sim}{R} \parallel \underset{\sim}{Q}^\alpha$. Now by the above theorem, for any two particles $\underset{\sim}{Q}^a$ and $\underset{\sim}{Q}^b$ there are observers

$\Psi(\hat{\underset{\sim}{Q}}^a)$ and $\Psi(\hat{\underset{\sim}{Q}}^b)$ such that $\Psi(\hat{\underset{\sim}{Q}}^a) \parallel \Psi(\hat{\underset{\sim}{R}})$ and $\Psi(\hat{\underset{\sim}{Q}}^b) \parallel \Psi(\hat{\underset{\sim}{R}})$ and so $\Psi(\hat{\underset{\sim}{Q}}^a) \parallel \Psi(\hat{\underset{\sim}{Q}}^b)$. □

We now consider compositions of reflections which lead to: either spacelike translations if we compose reflections in two mutually parallel particles; or to "pseudo-rotations" if we compose reflections in two distinct particles which coincide at some event. These mappings are discussed in more detail in the following section §7.5.

THEOREM 44 (Mapping of an Indexed Class of Parallels)
Let COL be a collinear set of particles, and let col be the corresponding set of events. Let $\{\underset{\sim}{Q}^\alpha : \alpha \text{ real}, \underset{\sim}{Q}^\alpha \in COL\}$ *and* $\{\underset{\sim}{W}^\beta : \beta \text{ real}, \underset{\sim}{W}^\beta \in COL\}$ *be indexed classes of divergent and convergent parallels such that, for some* $\underset{\sim}{Q}^a$ *and* $\underset{\sim}{W}^d$,

$$\underset{\sim}{Q}^a \simeq \underset{\sim}{W}^d$$

Let $\underset{\sim}{V} \in COL$ *be a particle such that either*

(i) $\underset{\sim}{V}$ *coincides with* $\underset{\sim}{Q}^a (\simeq \underset{\sim}{W}^d)$ *at some event, or*

(ii) $\underset{\sim}{V} \vee \underset{\sim}{Q}^a (\simeq \underset{\sim}{W}^d)$, *or*

(iii) $\underset{\sim}{V} \wedge \underset{\sim}{Q}^a (\simeq \underset{\sim}{W}^d)$.

Then there is a mapping

$$\phi : col \to col$$

such that

$$\phi(\underset{\sim}{Q}^a) \simeq \phi(\underset{\sim}{W}^d) \simeq \underset{\sim}{V}$$

and the indexed classes of parallels are mapped

into indexed classes of parallels of the same type. That is, the indexed class of divergent parallels $\{\underset{\sim}{Q}^{\alpha}: \alpha$ *real*, $\underset{\sim}{Q}^{\alpha} \in COL\}$ *is mapped onto an indexed class of divergent parallels* $\{\underset{\sim}{U}^{\alpha}: \alpha$ *real*, $\underset{\sim}{U}^{\alpha} \in COL\}$ *and, furthermore,*

$$\phi: [Q^x_y] \to [U^x_y] .$$

A similar result applies to the indexed class of convergent parallels $\{\underset{\sim}{W}^{\beta}: \beta$ *real*, $\underset{\sim}{W}^{\beta} \in COL\}$.

A mapping corresponding to case (i) is called a *"pseudo-rotation"*, while mappings corresponding to cases (ii) and (iii) are called *spacelike translations.*

This theorem is a consequence of Theorems 23 (§5.3), 39 (§7.3) and 43 (§7.4) together with its Corollary. The theorem is used in the proof of Theorems 45 (§7.4) and 48 (§7.5).

PROOF (See Fig. 53)
By Theorems 23 (§5.3) and 39 (§7.3), there is a particle $\underset{\sim}{T}$ midway between $\underset{\sim}{Q}^{a} (\simeq \underset{\sim}{W}^{d})$ and $\underset{\sim}{V}$. As in the previous corollary, we use the symbol Ψ to denote a reflection mapping in $\underset{\sim}{T}$, so that for any particle $\underset{\sim}{S} \in COL$,

$$(1) \qquad \left(f_{TS} \circ f_{ST}\right)^* = \left\{\left(f_{T\Psi(S)} \circ f_{\Psi(S)T}\right)^*\right\}^{-1} = f^+_{T\Psi(S)} \circ f^-_{\Psi(S)T}$$

(In this and all subsequent equations, record functions represent mappings between observers as explained in §3.6).

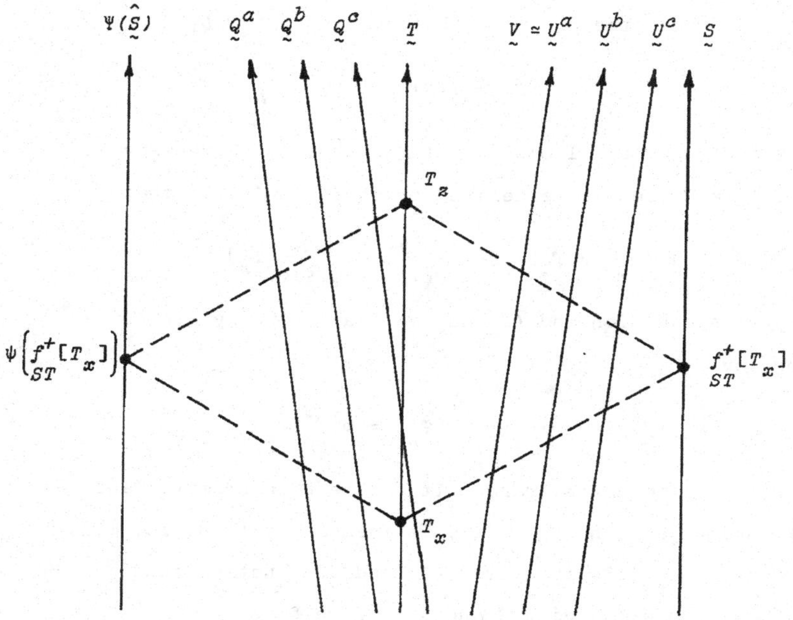

Fig. 53

We now define a mapping (denoted by lower case ψ)

$$\psi: \quad col \rightarrow col$$

such that, for any $T_x \in \underset{\sim}{T}$ and for any $\underset{\sim}{S} \in COL$,

$$\psi: \quad \underset{ST}{f^+}[T_x] \mapsto \underset{\psi(S)T}{f^-}[T_x] = \psi\left(\underset{ST}{f^+}[T_x]\right) ,$$

which clearly has the property that for any $\underset{\sim}{S} \in COL$,

$$\psi(\hat{\underset{\sim}{S}}) = \Psi(\hat{\underset{\sim}{S}}) ,$$

so from now on we shall only use the symbol ψ. Given any instant $T_x \in \underset{\sim}{T}$, let T_z be defined by equation (1) so that

$$[T_z] = f^-_{TS} \circ f^+_{ST}[T_x] = f^+_{T\psi(S)} \circ f^-_{\psi(S)T}[T_x] .$$

Then by this equation and (2) we see that, for any $\underset{\sim}{S} \in COL$, and for any $T_z \in \underset{\sim}{T}$,

(3) $$\psi : f^-_{ST}[T_z] \to f^+_{\psi(S)T}[T_z] = \psi\left(f^-_{ST}[T_z]\right) ,$$

since T_x and $\underset{\sim}{S}$ were arbitrary. It can be seen from (2) and (3) that ψ is a bijection and that ψ is idempotent (that is, ψ is its own inverse). So inverting (2) and (3), noting that $\psi = \psi^{-1}$, and interchanging $\underset{\sim}{S}$ and $\psi(\underset{\sim}{S})$, we have

(4) $$f^-_{TS} = f^+_{T\psi(S)} \circ \psi \quad \text{and} \quad f^+_{TS} = f^-_{T\psi(S)} \circ \psi .$$

By equations (2), (3) and (4) we see that for any observers $\hat{\underset{\sim}{R}}, \hat{\underset{\sim}{S}} \subset COL$,

(5) $$f^+_{RS} = f^+_{RT} \circ f^+_{TS} = \psi \circ f^-_{\psi(R)T} \circ f^-_{T\psi(S)} \circ \psi = \psi \circ f^-_{\psi(R)\psi(S)} \circ \psi$$

and similarly

(6) $$f^-_{RS} = \psi \circ f^+_{\psi(R)\psi(S)} \circ \psi .$$

Letting $\underset{\sim}{\hat{R}}$ and $\underset{\sim}{\hat{S}}$ be $\underset{\sim}{\hat{Q}}^c$ and $\underset{\sim}{\hat{Q}}^b$ respectively, we have

(7) $f^+_{Q^bQ^c} = \psi \circ f^-_{\psi(Q^b)\psi(Q^c)} \circ \psi$ and $f^-_{Q^bQ^c} = \psi \circ f^+_{\psi(Q^b)\psi(Q^c)} \circ \psi$

Right and left signal functions have been interchanged because ψ is a reflection mapping. In order that right and left signal functions should map onto right and left signal functions, respectively, we shall compose two reflection mappings. Thus we define a reflection mapping θ, which is a reflection in $\psi(\underset{\sim}{\hat{Q}}^a)$ $(= \underset{\sim}{\hat{V}})$ so that for any particle $\underset{\sim}{S} \in COL$ and for any instant $V_x \in \underset{\sim}{V}$,

(8) $\theta: \qquad col \to col$

$$f^+_{S\psi(Q^a)}(\psi[Q^a_x]) \mapsto f^-_{\theta(S)\psi(Q^a)}(\psi[Q^a_x]) = \theta\left[f^+_{S\psi(Q^a)}(\psi[Q^a_x])\right],$$

which is analogous to the definition (2). Consequently, by similar arguments, we find that

(9) $\theta: f^-_{S\psi(Q^a)}(\psi[Q^a_x]) \mapsto f^+_{\theta(S)\psi(Q^a)}(\psi[Q^a_x]) = \theta\left[f^-_{S\psi(Q^a)}(\psi[Q^a_x])\right]$

and

(10) $f^+_{RS} = \theta \circ f^-_{\theta(R)\theta(S)} \circ \theta$ and $f^-_{RS} = \theta \circ f^+_{\theta(R)\theta(S)} \circ \theta$.

If we now define

(11) $$\phi \overset{def}{=} \theta \circ \psi$$

and combine equations (7) and (10), we obtain

(12) $f^{+}_{Q^bQ^c} = \phi^{-1} \circ f^{+}_{\phi(Q^b)\phi(Q^c)} \circ \phi$ and $f^{-}_{Q^bQ^c} = \phi^{-1} \circ f^{-}_{\phi(Q^b)\phi(Q^c)} \circ \phi$,

since $\phi^{-1} = (\theta\circ\psi)^{-1} = \psi^{-1} \circ \theta^{-1} = \psi \circ \theta$. We have already noted that $\{\psi(\hat{\underset{\sim}{Q}}^{\alpha})\}$ is a class of parallels of the same type as $\{\hat{\underset{\sim}{Q}}^{\alpha}\}$, and similarly $\{\phi(\hat{\underset{\sim}{Q}}^{\alpha})\}$ is a class of parallels of the same type as $\{\hat{\underset{\sim}{Q}}^{\alpha}\}$. For each real α we take a particle $\underset{\sim}{U}^{\alpha} \in \phi(\hat{\underset{\sim}{Q}}^{\alpha})$, which is possible by the Axiom of Choice, and index the instants of this set of particles such that, for all real x and y,

(13) $$[U^x_y] = \phi[Q^x_y] .$$

Accordingly, equations (12) imply that

(14) $$[U^b_{x+b-c}] = f^{+}_{U^bU^c}[U^c_x] \quad \text{and} \quad [U^b_{x-b+c}] = f^{-}_{U^bU^c}[U^c_x] ,$$

and hence

(15) $$[U^b_{x-2b+2c}] = \left(f_{U^bU^c} \circ f_{U^cU^b}\right)^{*}[U^b_x] .$$

Equations (13), (14), and (15) show that with the indexing specified by (13), the set $\{\underset{\sim}{U}^{\alpha}\}$ is an indexed class of parallels, and by (2) and (8),

$$\phi(\hat{\underset{\sim}{Q}}^{a}) = \hat{\underset{\sim}{U}}^{a} = \hat{\underset{\sim}{V}} .$$

A similar result can be obtained for the class of convergent parallels $\{\underset{\sim}{W}^{\beta}\}$ by replacing $\hat{\underset{\sim}{Q}}^{c}$ and $\hat{\underset{\sim}{Q}}^{b}$ of equations (7) with $\hat{\underset{\sim}{W}}^{c}$ and $\hat{\underset{\sim}{W}}^{b}$, respectively, and then proceeding along similar lines. □

THEOREM 45 (Time Translation)

Let COL be a collinear set of particles and let $\{\underset{\sim}{Q}^{\alpha}: \alpha \text{ real}, \underset{\sim}{Q}^{\alpha} \in COL\}$ *and* $\{\underset{\sim}{R}^{\beta}: \beta \text{ real}, \underset{\sim}{R}^{\beta} \in COL\}$ *be indexed classes of divergent and convergent parallels, respectively, with*

$$\underset{\sim}{Q}^{0} \simeq \underset{\sim}{R}^{0} \quad \text{and} \quad Q_0^0 \simeq R_0^0 .$$

Let $Q_a^0 \in \underset{\sim}{Q}^0$ *with* $Q_0^0 < Q_a^0 < Q_1^0$. *There is a bijection*

$$\tau: \quad col \to col$$

$$[Q_y^{\alpha}] \mapsto [Q_{a+ky}^{k\alpha}]$$

$$[R_z^{\beta}] \mapsto [R_{b+\ell z}^{\ell\beta}] \;,$$

where k, ℓ, b *are positive real numbers. In particular*

$$\tau: \hat{\underset{\sim}{Q}}^{\alpha} \to \hat{\underset{\sim}{Q}}^{k\alpha} , \; \hat{\underset{\sim}{R}}^{\beta} \to \hat{\underset{\sim}{R}}^{\ell\beta} , \; [Q_y^0] \to [Q_{a+ky}^0] , \; [R_z^0] \to [R_{b+\ell z}^0] .$$

Furthermore, for any positive integer n,

$$Q^0_{a(1+k+\cdots+k^n)} \simeq R^0_{b(1+\ell+\cdots+\ell^n)} .$$

The mapping τ is called a *time translation mapping.*

This theorem is a consequence of Theorems 18 (§4.3), 33 (§6.4), 41 (§7.3) and 44 (§7.4). It is used in the proof of Theorems 46 (§7.5) and 48 (§7.5).

PROOF (see Fig. 54)

Since $Q^0_0 < Q^0_a < Q^0_1$, it follows that

$$Q^1_{-1} < Q^1_{a+1} \quad \text{and} \quad Q^1_{a-1} < Q^1_1 ,$$

so by Theorem 33 (§6.4) there is a particle $\underset{\sim}{T} \in COL$ such that

$$\text{(1)} \qquad \left(f_{Q^1 T} \circ f_{T Q^1} \right)^* (Q^1_{-1}) = Q^1_{a-1} \quad \text{and}$$

$$\text{(2)} \qquad \left(f_{Q^1 T} \circ f_{T Q^1} \right)^* (Q^1_{a+1}) = Q^1_1 .$$

We will now define an indexed class of parallels $\{\underset{\sim}{S}^\alpha\}$ (it is immaterial whether we choose divergent or convergent parallels) such that $\underset{\sim}{S}^1 \in \hat{\underset{\sim}{T}}$,

$$S^1_{-1} \overset{def}{=} f^+_{S^1 Q^1} (Q^1_{-1}) \quad \text{and} \quad S^1_1 \overset{def}{=} f^+_{S^1 Q^1} (Q^1_{a+1}) ;$$

whence from (1) and (2),

$$Q^1_{a-1} = f^-_{Q^1 S^1} (S^1_{-1}) \quad \text{and} \quad Q^1_1 = f^-_{Q^1 S^1} (S^1_1) ,$$

and so by Theorem 41 (7.3) and Theorem 18 (4.3)

$$f^+_{Q^2 S^1} (S^1_{-1}) = Q^2_0 \quad \text{and} \quad f^+_{Q^0 S^1} (S^1_1) = Q^0_a \quad \text{and}$$

$$f^-_{Q^0 S^1} (S^1_{-1}) = Q^0_a \quad \text{and} \quad f^-_{Q^2 S^1} (S^1_1) = Q^2_0 ;$$

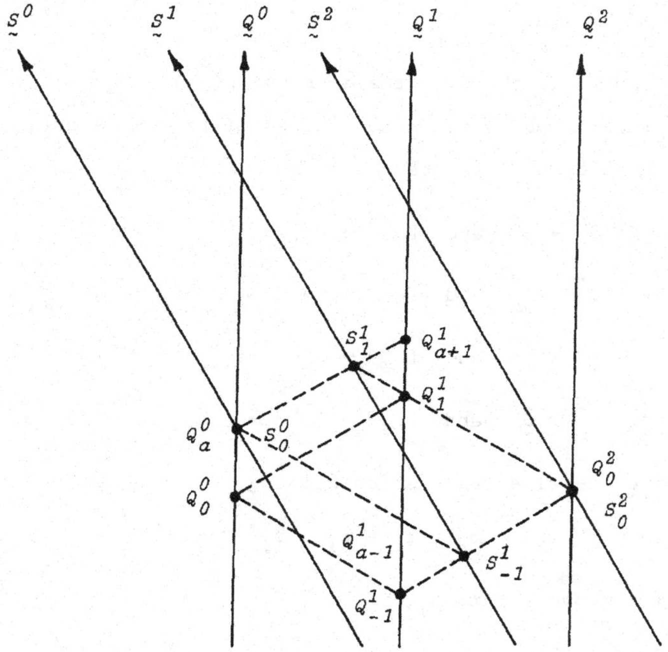

Fig. 54

these four relations imply that, if we define particles $\underset{\sim}{S}^0, \underset{\sim}{S}^2 \in \{\underset{\sim}{S}^\alpha\}$ in accordance with Theorem 41 (§7.3), then $\underset{\sim}{S}^0$ coincides with $\underset{\sim}{Q}^0$ at $[Q_a^0]$ (= $[S_0^0]$) and $\underset{\sim}{S}^2$ coincides with $\underset{\sim}{Q}^2$ at $[Q_0^2]$ (= $[S_0^2]$) .

By the previous theorem, there are mappings $\phi_1, \phi_2, \phi_3, \phi_4$: *col* → *col* which send classes of parallels onto classes of parallels of the same type such that

$$\phi_1\colon\ \hat{\underset{\sim}{Q}}^0 \to \hat{\underset{\sim}{Q}}^2 \quad \text{and} \quad [Q_0^0] \to [Q_0^2] \ ,$$

$$\phi_2\colon\ \hat{\underset{\sim}{Q}}^2 \to \hat{\underset{\sim}{S}}^2 \quad \text{and} \quad [Q_0^2] \to [S_0^2] = [Q_0^2] \ ,$$

$$\phi_3\colon\ \hat{\underset{\sim}{S}}^2 \to \hat{\underset{\sim}{S}}^0 \quad \text{and} \quad [S_0^2] \to [S_0^0] \quad \text{and}$$

$$\phi_4\colon\ \hat{\underset{\sim}{S}}^0 \to \hat{\underset{\sim}{Q}}^0 \quad \text{and} \quad [S_0^0] \to [Q_a^0] = [S_0^0] \ .$$

We now define the time translation mapping

$$\tau \overset{def}{=} \phi_4 \circ \phi_3 \circ \phi_2 \circ \phi_1$$

and so by the above relations,

$$\tau\colon\ \hat{\underset{\sim}{Q}}^0 \to \hat{\underset{\sim}{Q}}^0 \quad \text{and} \quad [Q_0^0] \to [Q_a^0] \ .$$

Also, by the previous theorem,

$$\tau\colon\ \{\hat{\underset{\sim}{Q}}^\alpha\} \to \{\hat{\underset{\sim}{Q}}^\alpha\} \ , \ \{\hat{\underset{\sim}{R}}^\beta\} \to \{\hat{\underset{\sim}{R}}^\beta\} \ .$$

However we can not assert that τ sends each parallel onto itself; we can only assert that τ sends the right side of $\underset{\sim}{Q}^0$

onto itself (and the left side of $\underset{\sim}{Q}^0$ onto itself). Accordingly, there are real positive numbers k and ℓ such that

$$\tau\colon \hat{\underset{\sim}{Q}}^1 \to \hat{\underset{\sim}{Q}}^k \quad , \quad \hat{\underset{\sim}{R}}^1 \to \hat{\underset{\sim}{R}}^\ell \quad ,$$

so by Theorem 42 (ii) (§7.3),

$$\tau\colon \hat{\underset{\sim}{Q}}^\alpha \to \hat{\underset{\sim}{Q}}^{k\alpha} \quad , \quad \hat{\underset{\sim}{R}}^\beta \to \hat{\underset{\sim}{R}}^{\ell\beta} \; .$$

Since $Q^0_x = f^-_{Q^0 Q^{x/2}} \circ f^+_{Q^{x/2} Q^0}(Q^0_0)$ and $R^0_y = f^-_{R^0 R^{y/2}} \circ f^+_{R^{y/2} R^0}(R^0_0)$,

it follows from the previous theorem that there is a positive real number b such that

$$\tau\colon [Q^0_x] \to [Q^0_{a+kx}] \; , \; [R^0_y] \to [R^0_{b+\ell y}] \; .$$

By Theorem 41 (§7.3) and the previous theorem,

$$\tau\colon [Q^\alpha_{x+\alpha}] = f^+_{Q^\alpha Q^0}[Q^0_x] \to [Q^{k\alpha}_{a+kx+k\alpha}] = f^+_{Q^{k\alpha} Q^0}[Q^0_{a+kx}]$$

and similar considerations apply to convergent parallels, so

$$\tau\colon [Q^\alpha_x] \to [Q^{k\alpha}_{a+kx}] \; , \; [R^\beta_y] \to [R^{\ell\beta}_{b+\ell y}] \; .$$

Since $Q^0_0 \simeq R^0_0$, it follows that for any positive integer n,

$$\tau^n\colon [Q^0_0] = [R^0_0] \to \left[Q^0_{a(1+k+\cdots+k^n)}\right] = \left[R^0_{b(1+\ell+\cdots+\ell^n)}\right] \; . \; \square$$

COROLLARY. *The time translation* τ *has an inverse*

$$\tau^{-1}: \quad col \to col$$

$$[Q_y^{\alpha}] \mapsto [Q_{-a/k\,+\,y/k}^{\alpha/k}]$$

$$[R_z^{\beta}] \mapsto [R_{-b/\ell\,+\,z/\ell}^{\beta/\ell}]$$

This corollary is used in the proof of the following theorem (Theorem 46 (§7.5)).

PROOF. The mapping τ^{-1} is clearly a bijection and it is easily verified that the composition of τ and τ^{-1}, in either order, is the identity mapping. □

§7.5 Linearity of Modified Signal Functions

The culmination of the present theory of parallelism is in the next theorem where we show that there is only one type of parallel. The axiom which implies the uniqueness of parallelism is the Signal Axiom (Axiom I, §2.2). It turns out that the only space-time satisfying our axioms is the Minkowski space-time, which shares the property of uniqueness of parallelism with the Euclidean geometry. Hyperbolic geometry, which is the other absolute geometry, does not have the property of uniqueness of parallelism; likewise the de Sitter space-time, which does not satisfy the Signal Axiom (Axiom I, §2.2) but which does satisfy appropriately modified versions of the remaining axioms, also does not have the property of uniqueness of parallelism.

The property of uniqueness of parallelism allows the space and time translation mappings of the previous section to be expressed relative to any indexed class of parallels (Theorems 47 and 48). The final theorem of this section involves the linearity of modified signal functions, which is the stage at which the geometrical discussion ceases temporarily and the kinematics begins.

THEOREM 46 ("Euclidean" Parallel Theorem)

Let COL be a collinear set of particles and let col be the corresponding set of events. Let $\underset{\sim}{Q}, \underset{\sim}{S} \in COL$ *and let* $[W_x] \in col$. *Then*

$$\underset{\sim}{S} \vee (\underset{\sim}{Q}, [W_x]) \Leftrightarrow \underset{\sim}{S} \wedge (\underset{\sim}{Q}, [W_x]) .$$

That is, given any particle and any event, there is exactly one parallel to the given particle through the given event .

This theorem is a consequence of Theorems 6(§2.9), 12 (§3.5), 19 (§4.3), 36 (§7.1), 40 (§7.3), 41 (§7.3) and its corollary, and 45 (§7.4). It is used in the proof of Theorems 47 (§7.5), 49 (§7.5) and 60 (§9.4).

PROOF. Let $\{\underset{\sim}{Q}^{\alpha}\}$ and $\{\underset{\sim}{R}^{\beta}\}$ be classes of divergent and convergent parallels as in the previous theorem with $\underset{\sim}{Q} \simeq \underset{\sim}{Q}^0 \simeq \underset{\sim}{R}^0$ and $Q_0^0 \simeq R_0^0$. Since the real positive numbers k, ℓ (of the previous theorem) are as yet unknown, we shall consider the following cases:
(i) $\ell = 1$, (i') $k = 1$, (ii) $k < 1$ and $\ell > 1$,
(ii') $k > 1$ and $\ell < 1$, (iii) $k > 1$ and $\ell > 1$,
(iii') $k < 1$ and $\ell < 1$. By the corollary to Theorem 41 (§7.3) it is sufficient to show that there are two parallels, one diverging from $\underset{\sim}{Q}$ and one converging to $\underset{\sim}{Q}$, which are permanently coincident.

By the previous theorem,

(1) $$k \neq 1 \Rightarrow \tau[Q^0_{a/(1-k)}] = [Q^0_{a/(1-k)}] \text{, and}$$

$$\ell \neq 1 \Rightarrow \tau[R^0_{b/(1-\ell)}] = [R^0_{b/(1-\ell)}] .$$

That is, the event(s)

$$[Q^0_{a/(1-k)}] \text{ and } [R^0_{b/(1-\ell)}]$$

are fixed with respect to the time translation τ. Since $Q^0_a \simeq R^0_b$, and

(2) $$\left(f_{Q^0Q^{a/2}} \circ f_{Q^{a/2}Q^0}\right)^*(Q^0_0) = Q^0_a \text{ and } \left(f_{R^0R^{b/2}} \circ f_{R^{b/2}R^0}\right)^*(R^0_0) = R^0_b ,$$

it follows by Theorem 19 (§4.3) that the particles $\underset{\sim}{Q}^{a/2}$ and $\underset{\sim}{R}^{b/2}$ coincide at the event

$$\left[f^{+}_{Q^{a/2}Q^{0}}(Q^{0}_{0})\right] .$$

Also by Theorem 41 (§7.3)

(3) $$\left(f_{Q^{0}Q^{ka/2}} \circ f_{Q^{ka/2}Q^{0}}\right)^{*}(Q^{0}_{a}) = Q^{0}_{a+ka} \quad \text{and}$$

$$\left(f_{R^{0}R^{\ell b/2}} \circ f_{R^{\ell b/2}R^{0}}\right)^{*}(R^{0}_{b}) = R^{0}_{b+\ell b} .$$

We shall now discuss each case separately.

Case (i) $\ell = 1$

By the previous theorem,

$$[Q^{0}_{a+ka}] = \tau[Q^{0}_{a}] = \tau[R^{0}_{b}] = [R^{0}_{2b}] .$$

If $k = 1$, Theorem 19 (§4.3) and equations (3) imply that the particles $\underset{\sim}{Q}^{a/2}$ and $\underset{\sim}{R}^{b/2}$ coincide at the event $\left[f^{+}_{Q^{a/2}Q^{0}}(Q^{0}_{a})\right]$.

But these two particles also coincide at the event $\left[f^{+}_{Q^{a/2}Q^{0}}(Q^{0}_{0})\right]$, and so by Theorem 6 (§2.9)

$$\underset{\sim}{Q}^{a/2} \simeq \underset{\sim}{R}^{b/2}$$

If $k \neq 1$, then since

$$\langle \underset{\sim}{Q}^{0}, \underset{\sim}{R}^{b/2}, \underset{\sim}{Q}^{a/2} \rangle \ \textit{after} \ \left[f^{+}_{Q^{a/2}Q^{0}}(Q^{0}_{0})\right] ,$$

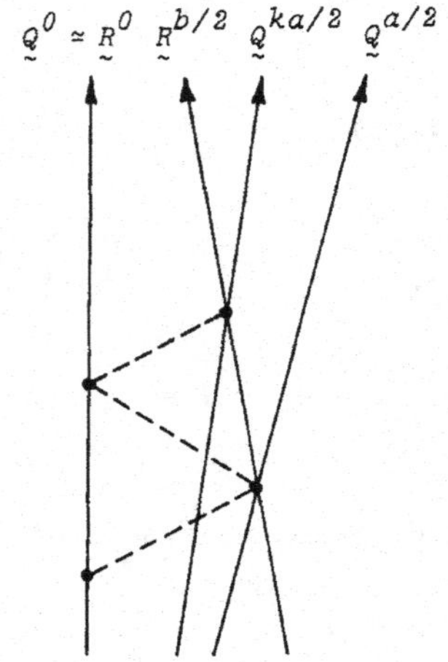

Fig. 55

it follows that $k < 1$ (see Fig. 55). Since $k \neq 1$, $[Q^0_{a/(1-k)}]$ is a fixed event. Applying the time translation mapping τ n times,

$$\tau^n\colon [Q^0_a] = [R^0_b] \to \left[Q^0_{a(1-k^{n+1})/(1-k)}\right] = \left[R^0_{b(n+1)}\right] .$$

Thus for every positive integer n,

$$R^0_{b(n+1)} \simeq Q^0_{a(1-k^{n+1})/(1-k)} < Q^0_{a/(1-k)} .$$

But by Theorem 12 (§3.5) this would imply that all members of $\{\underset{\sim}{R}^\beta\}$ coincide at some event before $[Q^0_{a/(1-k)}]$, which is a contradiction.

Case (i') $k = 1$

The proof is similar to the proof of case (i).

Case (ii) $k < 1$ and $\ell > 1$

Applying the time translation mapping τ n times as in case (i),

$$\tau^n: \quad [Q^0_a] = [R^0_b] \to \left[Q^0_{a(1-k^{n+1})/(1-k)}\right] = \left[R^0_{b(1+\ell+\cdots+\ell^n)}\right] .$$

Thus, for all integers n ,

$$R^0_{b(1+\ell+\cdots+\ell^n)} < Q^0_{a/(1-k)} ,$$

which is a contradiction since the sequence $\left\{R^0_{b(1+\ell+\cdots+\ell^n)} : n = 1,2,\ldots\right\}$ is unbounded.

Case (ii') $k > 1$ and $\ell < 1$

The proof is similar to case (ii) above.

Case (iii) $k > 1$ and $\ell > 1$

By (1) there are fixed events $[Q^0_{a/(1-k)}]$ and $[R^0_{b/(1-\ell)}]$; we will first show that these events are identical. By Theorem 36 (§7.1) there is a particle $\underset{\sim}{Q}^c$ such that

$$\left(f_{Q^0Q^c} \circ f_{Q^cQ^0}\right)^*[Q^0_{a/(1-k)}] = [R^0_{b/(1-\ell)}] .$$

By the previous theorem, the time translation τ maps this equation onto the equation

$$\left(f_{Q^0Q^{kc}} \circ f_{Q^{kc}Q^0}\right)^*[Q^0_{a/(1-k)}] = [R^0_{b/(1-\ell)}] .$$

and hence by Theorem 19 (§4.3) we conclude that $kc = c$, but $k \neq 1$, whence $c = 0$, so that

$$[Q^0_{a/(1-k)}] = [R^0_{b/(1-\ell)}] \quad .$$

We now define an affine transformation on the indexing of both classes of parallels by defining an indexed divergent class of parallels $\{\underset{\sim}{S}^\alpha : \alpha \text{ real}, \underset{\sim}{S}^\alpha \in COL\}$ and an indexed convergent class of parallels $\{\underset{\sim}{T}^\beta : \beta \text{ real}, \underset{\sim}{T}^\beta \in COL\}$ such that

$\underset{\sim}{S}^0 \simeq \underset{\sim}{Q}^0$, $\underset{\sim}{T}^0 \simeq \underset{\sim}{R}^0$, $S^0_0 \simeq Q^0_{a/(1-k)}$, $T^0_0 \simeq R^0_{b/(1-\ell)}$, and

$$f_{S^0 S^{1/2}} \circ f_{S^{1/2} S^0} (S^0_0) = S^0_1 \simeq T^0_1 = f_{T^0 T^{1/2}} \circ f_{T^{1/2} T^0} (T^0_0) \quad .$$

The time translation mapping τ has properties which can be expressed more simply with respect to the indexed classes $\{\underset{\sim}{S}^\alpha\}$ and $\{\underset{\sim}{T}^\beta\}$. Thus

$$\tau : \hat{\underset{\sim}{S}}^\alpha \to \hat{\underset{\sim}{S}}^{k\alpha} , \ \hat{\underset{\sim}{T}}^\beta \to \hat{\underset{\sim}{T}}^{\ell\beta} , \ [S^0_y] \to [S^0_{ky}] , \ [T^0_z] \to [T^0_{\ell z}] ,$$

and also, since $S^0_1 \simeq T^0_1$,

$$S^0_k \simeq T^0_\ell \quad .$$

Given any positive integer n , the time translation mapping τ can be applied n times and so

$$\tau^n : \hat{\underset{\sim}{S}}^\alpha \to \hat{\underset{\sim}{S}}^{k^n\alpha} , \ \hat{\underset{\sim}{T}}^\beta \to \hat{\underset{\sim}{T}}^{\ell^n\beta} , \ \left[S^0_y\right] \to \left[S^0_{k^n y}\right] , \ \left[T^0_z\right] \to \left[T^0_{\ell^n z}\right] \tag{4}$$

and

$$S^0_{k^n} \simeq T^0_{\ell^n} \quad .$$

By the corollary to Theorem 40 (§7.3) the set of instants of $\underset{\sim}{Q}$ is order-isomorphic to the sets of real numbers which index the instants of $\underset{\sim}{S}^0$ and $\underset{\sim}{T}^0$, so there is a strictly monotonic bijection g from the real indices of $\underset{\sim}{S}^0$ to the real indices of $\underset{\sim}{T}^0$; that is,

$$(5) \qquad g\colon \underset{\sim}{S}^0 \to \underset{\sim}{T}^0$$
$$y \mapsto z \textit{ if and only if } S^0_y \simeq T^0_z \quad .$$

We will now define a function h so that

$$(6) \qquad h(y) \overset{def}{=} g(y+1) - g(y)$$

We will show that h is an unbounded function by showing that, for each real number $\beta > 1$, $\underset{\sim}{S}^{1/2}$ crosses $\underset{\sim}{T}^{\beta}$; for, if $\underset{\sim}{S}^{1/2}$ crosses $\underset{\sim}{T}^{h(y)/2}$ at the event $\left[f_{S^{1/2}S^0}(S^0_y)\right]$ then by Theorem 19 (§4.3) and the Indexing Theorem (Theorem 41, §7.3),

$$f_{S^0S^{1/2}} \circ f_{S^{1/2}S^0}(S^0_y) = S^0_{y+1} \quad \text{and}$$

$$f_{T^0T^{h(y)/2}} \circ f_{T^{h(y)/2}T^0}(T^0_{g(y)}) = T^0_{g(y+1)}$$

Suppose the contrary; that is, suppose that $\underset{\sim}{S}^{1/2}$ does not cross all $\underset{\sim}{T}^{\beta}$ (with $\beta > 1$). Then there is a smallest real number $\gamma > 1$

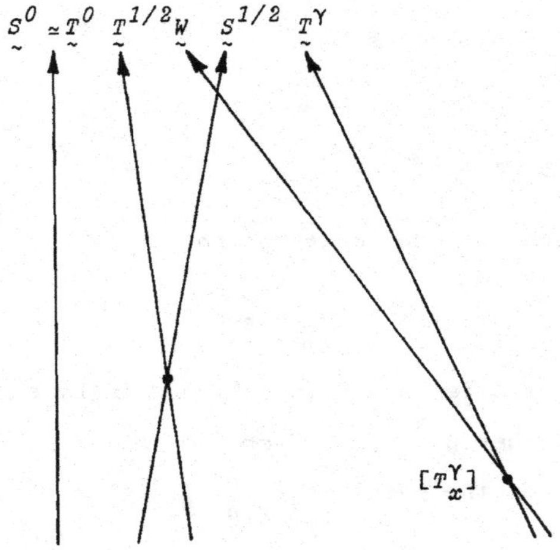
$\underset{\sim}{S}^0 \simeq \underset{\sim}{T}^0$
$\underset{\sim}{T}^{1/2}$
$\underset{\sim}{W}$
$\underset{\sim}{S}^{1/2}$
$\underset{\sim}{T}^{\gamma}$
$[T_x^{\gamma}]$

Fig. 56

such that $\underset{\sim}{S}^{1/2}$ does not cross $\underset{\sim}{T}^{\gamma}$ (see Fig. 56). Take any instant $T_x^{\gamma} \in \underset{\sim}{T}^{\gamma}$ and let $\underset{\sim}{W}$ be any particle which crosses $\underset{\sim}{T}^{\gamma}$ at $[T_x^{\gamma}]$ and which is to the left of $\underset{\sim}{T}^{\gamma}$ after $[T_x^{\gamma}]$. Then, for each real number $\beta < \gamma$, $\underset{\sim}{W}$ crosses the corresponding particle $\underset{\sim}{T}^{\beta}$; and hence $\underset{\sim}{W}$ crosses $\underset{\sim}{S}^{1/2}$ at some event after $[T_x^{\gamma}]$. Thus $\underset{\sim}{T}^{\gamma} \wedge \underset{\sim}{S}^{1/2}$ and so

$$\underset{\sim}{S}^{1/2} \wedge \underset{\sim}{T}^{\gamma} \;,\; \underset{\sim}{T}^{\gamma} \wedge \underset{\sim}{T}^{0} \;,\; \underset{\sim}{T}^{0} \simeq \underset{\sim}{S}^{0} \;,\; \underset{\sim}{S}^{0} \vee \underset{\sim}{S}^{1/2}$$

whence

$$\underset{\sim}{S}^{1/2} \wedge \underset{\sim}{S}^{0} \quad \text{and} \quad \underset{\sim}{S}^{0} \vee \underset{\sim}{S}^{1/2} \quad .$$

Consequently, by the corollary to Theorem 41 (§7.3), each divergent parallel is a convergent parallel and vice versa, in which case there is nothing further to prove, or h is unbounded. Furthermore, h must be a strictly monotonic increasing function, since otherwise $\underset{\sim}{S}^{1/2}$ would cross some convergent parallel at two distinct events, which would lead to a contradiction by Theorem 6 (§2.9). Thus, in the case where h is unbounded, there is some integer n_a such that:

(7) *for all real* $y > n_a$, $h(y) > 2$.

Since $k > 1$, the two sequences $(k^m\colon m=1,2,\cdots)$ and $(k^{m+1}-k^m\colon m=1,2,\cdots)$ are both unbounded so there is some integer n_b such that:

(8) *for all integers* $m > n_b$, $k^m > n_a$ and $k^{m+1} - k^m > 2$.

Let $n \stackrel{def}{=} max\{n_a, n_b\}$. Let $K(m)$ be the largest non-negative integer such that

(9) $$K(m) < k^{m+1} - k^m .$$

Then for any integer $m > n$,

(10) $$\frac{k^{m+1}-k^m}{2} < K(m) < k^{m+1} - k^m$$

By (4) and (5), for $m > n$,

$$\ell^{m+1} = g(k^{m+1})$$

$$> g(k^m+K(m))$$

by (9), since g is a strictly monotonic increasing function, and from (6)

$$\ell^{m+1} > h(k^m+K(m)-1) + h(k^m+K(m)-2) + \cdots + h(K^m) + g(k^m) .$$

Also (7) and (8) imply that $h(k^m) > 2$ and h is a strictly monotonic increasing function, so

$$\ell^{m+1} > 2K(m) + g(k^m)$$

$$> k^{m+1} - k^m + \ell^m \quad \text{by (4), (5) and (10) .}$$

Therefore, for any integer $m > n$,

$$\ell^{m+1} - \ell^m > k^{m+1} - k^m$$

and since $k > 1$ and $\ell > 1$,

$$\left(\frac{\ell}{k}\right)^m > \frac{k-1}{\ell-1}$$

which implies that

(11) $$k < \ell \quad .$$

We now establish an inequality which is opposite to (11) by a similar procedure. We define classes of divergent and convergent parallels $\{\underset{\sim}{U}^\alpha\}$ and $\{\underset{\sim}{V}^\beta\}$, respectively, which are not indexed classes of parallels in the sense of Theorem 41 (§7.3), because the indices of $\{\underset{\sim}{U}^\alpha\}$ and $\{\underset{\sim}{V}^\beta\}$ are *time reversed;* that is, for all α and β ,

$$b < c \Longleftrightarrow U^\alpha_b > U^\alpha_c \quad \text{and} \quad V^\beta_b > V^\beta_c \quad .$$

The relations corresponding to the relations of Theorem 41 (§7.3) are:

$$f^+_{U^cU^b}(U^b_a) = U^c_{a+b-c} \quad \text{and} \quad f^-_{U^cU^b}(U^b_a) = U^c_{a-b+c} \quad ,$$

whence

$$\left(f_{U^bU^c} \circ f_{U^cU^b}\right)^*(U^b_a) = U^b_{a+2b-2c}$$

and similar relations apply for the convergent class of parallels $\{\underset{\sim}{V}^\beta\}$. As before the classes of parallels $\{\underset{\sim}{U}^\alpha\}$ and $\{\underset{\sim}{V}^\beta\}$ are defined such that:

$$\underset{\sim}{U}^0 \simeq \underset{\sim}{Q}^0 , \underset{\sim}{V}^0 \simeq \underset{\sim}{R}^0 , U^0_0 \simeq Q^0_{a/(1-k)} , V^0_0 \simeq R^0_{b/(1-\ell)} \text{ and}$$

$$f^{-1}_{U^{1/2}U^0} \circ f^{-1}_{U^0U^{1/2}}(U^0_0) = U^0_1 \simeq V^0_1 = f^{-1}_{V^{1/2}V^0} \circ f^{-1}_{V^0V^{1/2}}(V^0_0) .$$

As before, the time reversal mapping τ can be applied n times and so

(4') $$\tau^n\colon \underset{\sim}{\hat{U}}^{\alpha} \to \underset{\sim}{\hat{U}}^{k^n\alpha} , \underset{\sim}{\hat{V}}^{\beta} \to \underset{\sim}{\hat{V}}^{\ell^n\beta} , \left[U^0_z\right] \to \left[U^0_{k^n z}\right] , \left[V^0_y\right] \to \left[V^0_{\ell^n y}\right]$$

and

$$U^0_{k^n} \simeq V^0_{\ell^n} .$$

Note that the subscript symbols y and z now apply to the indices of convergent and divergent classes of parallels, respectively, whereas before they applied in the opposite way. As before, there is a function

(5') $$G\colon \underset{\sim}{V}^0 \to \underset{\sim}{U}^0$$
$$y \mapsto z \text{ if and only if } V^0_y \simeq U^0_z .$$

Similarly we define

(6') $$H(y) \overset{def}{=} G(y+1) - G(y) .$$

As before

$$k^{m+1} = G(\ell^{m+1})$$

and so on, whence

(11') $$\ell < k \,,$$

which is a contradiction of (11).

Case (iii') $k < 1$ and $\ell < 1$

The proof is similar to the proof of case (iii).

We have seen that the only permissible combinations of k and ℓ are:
(i) $k = \ell = 1$, (iii) $k > 1$ and $\ell > 1$, and (iii') $k < 1$ and $\ell < 1$; and in each permissible case there is only one class of parallels, which are both divergent and convergent. Thus we have demonstrated the "uniqueness of parallelism". □

An immediate consequence of the previous theorem and Theorem 42 (§7.3) is that a time scale of any particle, defined relative to a class of parallels containing the given particle, is determined to within an arbitrary strictly increasing linear transformation and is called a *natural time scale*.

THEOREM 47 (Space Displacement Mapping)
Let COL be a collinear set of particles. Let $\{\underset{\sim}{Q}^{\alpha}: \alpha \text{ real}, \underset{\sim}{Q}^{\alpha} \in COL\}$ *be an indexed class of parallels. Given any real numbers a and b, there is a bijection*

$$\delta: col \rightarrow col$$
$$[Q_t^x] \mapsto [Q_t^{x-a+b}]$$

and for all $\underset{\sim}{R} \in COL$,

$$\delta(\hat{\underset{\sim}{R}}) \parallel \hat{\underset{\sim}{R}} \ .$$

Furthermore, for any indexed class of parallels $\{\underset{\sim}{U}^{\alpha}: \alpha \text{ real}, \underset{\sim}{U}^{\alpha} \in COL\}$, *there are real constants c and d such that*

$$\delta: [U_t^x] \rightarrow [U_{d+t}^{c+x}] \ .$$

The mapping δ is called a *space displacement mapping.*

This theorem is a consequence of Theorems 22 (§5.3), 39 (§7.3), 41 (§7.3), 42 (§7.3) and 46 (§7.5). It is used in the proof of Theorems 48 (§7.5) and 49 (§7.5).

PROOF. The case $a = b$ is trivial, so from now on we assume that $a \neq b$. This proof is based on the proof of Theorem 44 (§7.4) to which we shall constantly refer. Accordingly, we let $\underset{\sim}{V} \overset{def}{=} \underset{\sim}{Q}^b$ and we let $\underset{\sim}{T} \overset{def}{=} \underset{\sim}{Q}^{(a+b)/2}$.

Thus

$$\psi: [Q_t^x] \mapsto [Q_t^{a+b-x}]$$

and

$$\theta: [Q_t^x] \mapsto [Q_t^{2b-x}] ,$$

whence, if we define $\delta \overset{def}{=} \phi \overset{def}{=} \theta \circ \psi$,

(1) $$\delta: [Q_t^x] \mapsto [Q_t^{x-a+b}].$$

Given any particle $\underset{\sim}{R} \in COL$, if $a \neq b$, (1) implies that there is no event at which $\underset{\sim}{R}$ and $\delta(\underset{\sim}{\hat{R}})$ coincide, so by the previous theorem, $\delta(\underset{\sim}{\hat{R}}) \parallel \underset{\sim}{\hat{R}}$. If $a = b$, the space displacement δ is trivial and $\delta(\underset{\sim}{\hat{R}}) = \underset{\sim}{\hat{R}}$, so $\delta(\underset{\sim}{\hat{R}}) \parallel \underset{\sim}{\hat{R}}$ trivially. We conclude that for all $\underset{\sim}{R} \in COL$,

(2) $$\delta(\underset{\sim}{\hat{R}}) \parallel \underset{\sim}{\hat{R}} .$$

Theorem 41 (§7.3) implies that for any real numbers t,x,y with $x < y$,

(3) $$[U_t^x] \ \sigma \ [U_{t-x+y}^y] \quad \text{and} \quad [U_t^y] \ \sigma \ [U_{t+x-y}^x] .$$

These relations correspond to right and left modified signal functions respectively. We will now show that

(4) $$(\delta[U_t^x]) \ \sigma \ (\delta[U_{t-x+y}^y]) \quad \text{and} \quad (\delta[U_t^y]) \ \sigma \ (\delta[U_{t+x-y}^x]) .$$

Given particular real numbers t,x,y there are real numbers A,B,C,D such that

(5) $$[Q^A_B] = [U^x_t] \text{ and } [Q^C_D] = [U^y_{t-x+y}] ,$$

since $\{\underset{\sim}{Q}^{\alpha}\}$ is an indexed class of parallels. Now (3) implies that

$$[Q^A_B] \;\sigma\; [Q^C_D]$$

from which

(6) $$D = B - A + C .$$

Also by (1),

(7) $$\delta\colon [Q^A_B] \rightarrow [Q^{A-a+b}_B] , \; [Q^C_D] \rightarrow [Q^{C-a+b}_D] .$$

Consequently by Theorem 41 (§7.3) together with (6) and (7),

$$(\delta[Q^A_B]) \;\sigma\; (\delta[Q^C_D]) ,$$

and by (5) this is equivalent to the first relation of (4): the second relation can be proved similarly.

The relations (4) are in accordance with the results of Theorem 41 (§7.3) so we can define an indexed class of parallels $\{\underset{\sim}{V}^{\alpha}\colon \alpha \; real , \; \underset{\sim}{V}^{\alpha} \in COL\}$ such that

$$\delta[U^x_t] \overset{def}{=} [V^x_t] .$$

By (2), there is some $\underset{\sim}{V}^{\beta}$ such that

$$\underset{\sim}{U}^{0} \simeq \underset{\sim}{V}^{\beta} ,$$

and so by Theorem 44 (§7.3) there are real constants c, d, k such that for all real t and x,

$$(8) \qquad [U^{c+kx}_{d+kt}] = [V^{x}_{t}] \quad (= \delta[U^{x}_{t}]) .$$

If $k \neq 1$ we can choose $t = (|c|-d)/(k-1)$, then

$$([U^{0}_{t}]) \ \sigma \ (\delta[U^{0}_{t}]) ,$$

which is a contradiction by (1), since δ is a space displacement mapping, so $k = 1$; whence (8) becomes

$$\delta[U^{x}_{t}] = [U^{c+x}_{d+t}] . \quad \square$$

THEOREM 48 (Time Translation Mapping)
Let COL be a collinear set of particles.
Let $\{\underset{\sim}{U}^{\alpha} : \alpha \text{ real} , \underset{\sim}{U}^{\alpha} \in COL\}$ *be an indexed class of parallels with instants* $U^{b}_{c}, U^{b}_{d} \in \underset{\sim}{U}^{b}$ *such that* $U^{b}_{c} < U^{b}_{d}$. *There is a bijection*

$$\begin{aligned} \tau : col &\rightarrow col \\ [U^{b}_{t}] &\mapsto [U^{b}_{t-c+d}] . \end{aligned}$$

The mapping τ is called a *time displacement mapping.*

This theorem is a consequence of Theorems 18 (§4.3), 33 (§6.4), 41 (§7.3), 42 (§7.3), 44 (§7.4), 45 (§7.4) and 47 (§7.5). It is used in the proofs of Theorems 49 (§7.5) and 61 (§9.5).

PROOF. This proof is based on the proof of Theorem 45 (§7.4). We define an indexed class of parallels

$$\{\underset{\sim}{Q}^{\alpha} : \alpha \ real \ , \ \underset{\sim}{Q}^{\alpha} \ \varepsilon \ COL\}$$

such that, for some real number a with $0 < a < 1$,

$$\underset{\sim}{Q}^{0} \overset{def}{=} \underset{\sim}{U}^{b} \ , \ Q^{0}_{0} \overset{def}{=} U^{b}_{c} \ , \ Q^{0}_{a} \overset{def}{=} U^{b}_{d} \ .$$

By Theorem 42 (§7.3),

(1) $$[Q^{x}_{t}] = [U^{b+kx}_{c+kt}] \text{ where } k = \frac{d-c}{a} \ .$$

The space displacement mappings ϕ_1 and ϕ_3 (of the proof of Theorem 45 (§7.4)) are such that

$$\phi_1 : [Q^{0}_{0}] \rightarrow [Q^{2}_{0}] \quad \text{and} \quad \phi_3 : [Q^{2}_{0}] \rightarrow [Q^{0}_{a}] \ ,$$

so by the previous theorem,

$$\phi_1 : [Q^{x}_{t}] \rightarrow [Q^{2+x}_{t}] \text{ and } \phi_3 : [Q^{x}_{t}] \rightarrow [Q^{-2+x}_{t+a}] \ .$$

We now define a *time displacement mapping* $\tau^* \overset{def}{=} \phi_3 \circ \phi_1$, and so

$$\tau^* : [Q^{x}_{t}] \mapsto [Q^{x}_{t+a}] \ ,$$

whence from (1),

$$\tau^* : [U^{b+kx}_{c+kt}] \rightarrow [U^{b+kx}_{d+kt}] \ ;$$

which is equivalent to

$$\tau^*: [U^z_y] \rightarrow [U^z_{y-c+d}] ,$$

which is the required mapping. □

THEOREM 49 (Linearity of Modified Signal Functions)

If $\underset{\sim}{Q}$ and $\underset{\sim}{R}$ are particles in COL with natural time scales, then

$$f^+_{QR}(R_t) = Q_{at+b} \quad \text{and} \quad f^-_{QR}(R_t) = Q_{ct+d} ,$$

where a, b, c, d are constants and both a and c are positive. Furthermore

$$f^+_{RQ}(Q_t) = R_{(t-b)/a} \quad \text{and} \quad f^-_{RQ}(Q_t) = Q_{(t-d)/c} .$$

This theorem is a consequence of Theorems 9 (§3.2), 18 (§4.3), 41 (§7.3), 42 (§7.3), 46 (§7.5), 47 (§7.5) and 48 (§7.5). It is used in the proof of Theorems 50 (§8.1), 51 (§8.1) and 57 (§9.1).

PROOF. If $\underset{\sim}{Q}$ and $\underset{\sim}{R}$ coincide at no event, or if they are permanently coincident, they are parallel and the result is a special case of Theorem 42 (§7.3). Otherwise $\underset{\sim}{Q}$ and $\underset{\sim}{R}$ coincide at some event by Theorem 46 (§7.5).

We now define indexed classes of parallels $\{\underset{\sim}{S}^\alpha: \alpha \text{ real}, \underset{\sim}{S}^\alpha \in COL\}$ and $\{\underset{\sim}{U}^\alpha: \alpha \text{ real}, \underset{\sim}{U}^\alpha \in COL\}$

such that

$$\underset{\sim}{S}^0 \simeq \underset{\sim}{Q} \,, \quad \underset{\sim}{U}^0 \simeq \underset{\sim}{R} \quad \text{and} \quad S_0^0 \simeq Q_0^0 \,.$$

For any real number a, $\underset{\sim}{S}^a \not\parallel \underset{\sim}{U}^0$, so $\underset{\sim}{S}^a$ and $\underset{\sim}{U}^0$ coincide at some event by Theorem 46 (§7.5); that is, there are real numbers b and c such that

(1) $$[S_b^a] = [U_c^0] \,.$$

Let δ and τ be space and time translations, respectively, as in the preceding two theorems, such that

$$\delta\colon [S_t^x] \to [S_t^{x+a}] \quad \text{and} \quad \tau\colon [S_t^x] \to [S_{t+b}^x] \,.$$

Consequently we can define a mapping

$$\lambda \overset{def}{=} \delta \circ \tau = \tau \circ \delta$$

such that

(2) $$\lambda\colon [S_t^x] \to [S_{t+b}^{x+a}] \,.$$

Since τ is a composition of two displacement mappings, λ is a composition of three displacement mappings, so by Theorem 47 (§7.5) and (1),

(3) $$\lambda\colon [U_t^x] \to [U_{t+c}^x] \,.$$

Since δ and τ are bijections, λ is a bijection and so, for any integer n,

$$\lambda^n : [S^0_0] = [U^0_0] \rightarrow [S^{na}_{nb}] = [U^0_{nc}] .$$

Now a was arbitrary, so if we choose any positive integer m and substitute $a/2^m$ for a wherever a appears, we find that for any positive integer m and for any integer n,

$$\left[S^{an/2^m}_{bn/2^m}\right] = \left[U^0_{cn/2^m}\right] ;$$

that is, for any dyadic number p,

$$(4) \qquad [S^{ap}_{bp}] = [U^0_{cp}] .$$

Consequently by Theorem 41 (§7.3)

$$\underset{S^0U^0}{f^+} [S^0_{bp-ap}] = [U^0_{cp}] \quad \text{and} \quad \underset{S^0U^0}{f^-} [S^0_{bp+ap}] = [U^0_{cp}]$$

and, since signal functions are continuous by Theorem 9 (§3.2), it follows that for all real t,

$$\underset{S^0U^0}{f^+} [S^0_t] = [U^0_{ct/(b-a)}] \quad \text{and} \quad \underset{S^0U^0}{f^-} [S^0_t] = [U^0_{ct/(b+a)}] .$$

That is, the modified signal functions $\underset{S^0U^0}{f^+}$ and $\underset{S^0U^0}{f^-}$ are linear strictly increasing functions and therefore Theorem 42 (§7.3) implies that $\underset{QR}{f^+}$ and $\underset{QR}{f^-}$ are linear strictly increasing functions which can be written in the general form

$$f^{+}_{QR}(R_t) = Q_{At+B} \quad \text{and} \quad f^{-}_{QR}(R_t) = Q_{Ct+D} ,$$

where A,B,C,D are constants and A and C are positive. By Theorem 18 (§4.3)

$$f^{+}_{RQ}(Q_t) = R_{(t-B)/A} \quad \text{and} \quad f^{-}_{RQ}(Q_t) = R_{(t-D)/C} . \qquad \square$$

CHAPTER 8

ONE-DIMENSIONAL KINEMATICS

In this chapter all particles have natural time scales and modified signal functions are linear. We will often delete the particle symbol where there is no chance of ambiguity; for example, in the next theorem, instead of writing

$$\left(\underset{QS}{f} \circ \underset{SQ}{f}\right)^{*} Q_x = Q_{M_{SQ}(x-q)+q} \quad ,$$

we shall write

$$\left(\underset{QS}{f} \circ \underset{SQ}{f}\right)^{*}(x) = M_{SQ}(x-q)+q \quad .$$

§8.1 Rapidity is a Natural Measure for Speed

In this section we define "rapidity" which is a non-dimensional measure of speed. For collinear sets of particles, directed rapidities are composed by simple arithmetic addition, which means that rapidity is a natural measure for speed. The name "rapidity" is due to Robb [1921] who introduced this concept in a different way.

THEOREM 50. *Let* $\underset{\sim}{Q},\underset{\sim}{S},\underset{\sim}{T} \in COL$.

(i) *If* $\underset{\sim}{S} \nparallel \underset{\sim}{Q}$ *, there is a positive "constant of the motion"* M_{QS} *and a real number* q *such that*

$$\left(f_{QS} \circ f_{SQ}\right)^*(x) = M_{QS}(x-q)+q \ ,$$

where the real number q *is such that* $\underset{\sim}{S}$ *coincides with* $\underset{\sim}{Q}$ *at* $[Q_q]$.

(ii) $$M_{SQ} = (M_{QS})^{-1}$$

(iii) *If* $\underset{\sim}{R} \parallel \underset{\sim}{Q}$ *and* $\underset{\sim}{T} \parallel \underset{\sim}{S}$ *, then*

$$M_{RT} = M_{QS} \ .$$

REMARK. The constant of the motion M is invariant with respect to affine transformations of natural time scales, by part (iii).

This theorem is a consequence of Theorems 12 (§3.5), 18 (§4.3) and the previous theorem. It is used in the proof of Theorems 51 (§8.2) and 56 (§8.4).

PROOF. (i) By the previous theorem, both f^+_{SQ} and f^-_{QS} are linear functions, so their composition $\left(f_{QS} \circ f_{SQ}\right)^*$ is a linear function. If $\underset{\sim}{Q} \nparallel \underset{\sim}{S}$, there is some instant $Q_q \in \underset{\sim}{Q}$ such that $\underset{\sim}{S}$ coincides with $\underset{\sim}{Q}$ at $[Q_q]$, and so the record function is of the form

$$\left(f_{QS} \circ f_{SQ}\right)^*(x) = M_{QS}(x-q)+q \ ,$$

where $M_{QS} > 0$, by Theorem 12 (§3.5).

(ii) The previous theorem implies that, for any two particles $\underset{\sim}{Q}, \underset{\sim}{S} \in COL$, there are constants $\beta_{QS}, q_s^-, \alpha_{SQ}, s_q^+$ such that

$$\underset{QS}{f^-}(x) = \beta_{QS}x + q_S^- \quad \text{and} \quad \underset{SQ}{f^+}(x) = \alpha_{SQ}x + s_q^+ ,$$

so by Theorem 18 (§4.3),

$$\underset{SQ}{f^-}(x) = \beta_{QS}^{-1}x - \beta_{QS}^{-1}q_S^- \quad \text{and} \quad \underset{QS}{f^+}(x) = \alpha_{SQ}^{-1}x - \alpha_{SQ}^{-1}s_q^+ ;$$

that is

$$\beta_{SQ} = \beta_{QS}^{-1} \qquad \text{and} \qquad \alpha_{QS} = \alpha_{SQ}^{-1} ,$$

whence

$$M_{QS} = \beta_{SQ}\alpha_{SQ} = (\beta_{QS}\alpha_{SQ})^{-1} = M_{SQ}^{-1} .$$

(iii) The previous theorem implies that there are constants $\alpha_{QR}, \beta_{RQ}, r_q^-, q_r^+$ such that

$$\underset{RQ}{f^-}(x) = \beta_{RQ}x + r_q^- \qquad \text{and} \quad \underset{QR}{f^+}(x) = \alpha_{QR}x + q_r^+ ,$$

and since $\underset{\sim}{Q} \parallel \underset{\sim}{R}$,

$$\alpha_{QR}\beta_{RQ} = 1$$

so

$$\underset{RQ}{f^-}(x) = \alpha_{QR}^{-1}x + r_q^- \qquad \text{and} \quad \underset{QR}{f^+}(x) = \alpha_{QR}x + q_r^+ .$$

Consequently by Theorem 18 (§4.3),

$$\begin{aligned}\left(f_{RS} \circ f_{SR}\right)^*(x) &= f^-_{RS} \circ f^+_{SR}(x) \\ &= f^-_{RQ} \circ f^-_{QS} \circ f^+_{SQ} \circ f^+_{QR}(x) \\ &= f^-_{RQ} \circ \left(f_{QS} \circ f_{SQ}\right)^* \circ f^+_{QR}(x) \\ &= M_{QS}x + \alpha^{-1}_{QR}\left(M_{QS}(q^+_r - q) + q\right) + r^-_q .\end{aligned}$$

Thus we have shown that

$$M_{RS} = M_{QS} ,$$

and similarly, since $\underset{\sim}{T} \parallel \underset{\sim}{S}$,

$$M_{TR} = M_{SR} ,$$

and so by (ii),

$$M_{RT} = (M_{TR})^{-1} = (M_{SR})^{-1} = M_{RS} = M_{QS} . \qquad \square$$

Given any particles $\underset{\sim}{Q}, \underset{\sim}{S} \in COL$ such that $\underset{\sim}{S} \parallel \underset{\sim}{Q}$, Theorem 42 (§7.3) shows that

$$\left(f_{QS} \circ f_{SQ}\right)^*(x) = x + 2d ,$$

where d is a real constant. The constant of the motion M_{SQ} is 1 , and is therefore not shown explicitly. The results of the preceding theorem (with the exception of (i)) apply trivially to the case where $\underset{\sim}{S} \parallel \underset{\sim}{Q}$.

Given any two particles $\underset{\sim}{Q}, \underset{\sim}{S} \in COL$ we define the *directed rapidity* of $\underset{\sim}{S}$ relative to $\underset{\sim}{Q}$ to be

$$r_{SQ} \overset{def}{=} \tfrac{1}{2} \log_e M_{SQ} .$$

Since $M > 0$,

$$-\infty < r < \infty .$$

In the case of parallel particles, $M = 1$ and hence $r = 0$. We call $|r_{SQ}|$ the *relative rapidity* of $\underset{\sim}{S}$ with respect to $\underset{\sim}{Q}$.

The next theorem is in no way surprising, since rapidity was defined so that we would have a measure for speed which is unbounded and which is composed by simple arithmetic addition. Also, rapidities are unaltered by arbitrary affine transformations of natural time scales, so the following result is invariant with respect to transformations of natural time scales.

THEOREM 51 (Addition Law for Directed Rapidity)
Given $\underset{\sim}{Q}, \underset{\sim}{S}, \underset{\sim}{T} \in COL$,

$$r_{QT} = r_{QS} + r_{ST} .$$

That is, "rapidity is a natural measure for speed".

This theorem is a consequence of Theorems 18 (§4.3), 49 (§7.5) and 50 (§8.1). It is used in the proof of Theorems 52 (§8.2) and 57 (§9.1).

PROOF. By Theorem 49 (§7.5) there are real constants $\beta_{QS}, q_s^-; \alpha_{SQ}, s_q^+; \beta_{ST}, s_t^-; \alpha_{TS}, t_s^+$ such that

$$f_{QS}^-(x) = \beta_{QS}x + q_s^- \quad \text{and} \quad f_{SQ}^+(x) = \alpha_{SQ}x + s_q^+ \quad \text{and}$$

$$f_{ST}^-(x) = \beta_{ST}x + s_t^- \quad \text{and} \quad f_{TS}^+(x) = \alpha_{TS}x + t_s^+ \quad .$$

By Theorem 18 (§4.3),

$$f_{QT}^-(x) = \beta_{QS}\beta_{ST}x + \beta_{QS}s_t^- + q_s^- \quad \text{and}$$

$$f_{TQ}^+(x) = \alpha_{TS}\alpha_{SQ}x + \alpha_{TS}s_q^+ + t_s^+ ,$$

whence, as in part (ii) of the previous theorem,

$$\begin{aligned} M_{QT} &= \beta_{QS}\beta_{ST}\alpha_{TS}\alpha_{SQ} \\ &= \beta_{QS}\alpha_{SQ}\beta_{ST}\alpha_{TS} \\ &= M_{QS}M_{ST} \quad , \end{aligned}$$

and taking logarithms of both sides,

$$r_{QT} = r_{QS} + r_{ST} \quad . \qquad \square$$

COROLLARY (Urquhart's Theorem : see Szekeres [1968])
Given $\underset{\sim}{Q}, \underset{\sim}{S} \in COL$,

$$r_{QS} = -r_{SQ} \quad .$$

PROOF. Put $\underset{\sim}{T} = \underset{\sim}{Q}$ in the above theorem. □

If $\{\underset{\sim}{Q}^{\alpha}: \alpha \text{ real}, \underset{\sim}{Q}^{\alpha} \in COL\}$ and $\{\underset{\sim}{S}^{\beta}: \beta \text{ real}, \underset{\sim}{S}^{\beta} \in COL\}$ are two classes of parallels, it follows from Theorem 50 (iii), that there is a unique directed rapidity of the class of parallels $\{\underset{\sim}{Q}^{\alpha}\}$ relative to the class of parallels $\{\underset{\sim}{S}^{\beta}\}$.

§8.2 Congruence of a Collinear Set of Particles

Given any two particles $\underset{\sim}{Q}, \underset{\sim}{S} \in COL$, there are constants $\alpha_{SQ}, \beta_{QS}, s_q^+, q_s^-$ such that

$$f^+_{SQ}(x) = \alpha_{SQ}x + s_q^+ \quad \text{and} \quad f^-_{QS}(x) = \beta_{QS}x + q_s^- \; ;$$

we say that $\underset{\sim}{Q}$ and $\underset{\sim}{S}$ are *congruent* if

$$\alpha_{SQ} = \beta_{QS} \; ,$$

which is equivalent to the condition

$$\alpha_{QS} = \beta_{SQ} \; ,$$

since $\alpha_{QS} = \alpha_{SQ}^{-1}$ and $\beta_{SQ} = \beta_{QS}^{-1}$.

The word "congruent" has also been used by Milne [1948] in a different sense: here, we use the work "synchronous" (see the following §8.3) where Milne used the work "congruent". We have departed in terminology because the word congruent is very descriptive of the idea of equality of time durations.

THEOREM 52

Congruence is an equivalence relation on a collinear set of particles.

This theorem is a consequence of Theorem 18 (§4.3) and is used in the proof of Theorem 53 (§8.2).

PROOF. By definition, congruence is a reflexive and symmetric relation.

In order to show that congruence is a transitive relation, we consider three particles $\underset{\sim}{Q}, \underset{\sim}{R}, \underset{\sim}{S} \in COL$ such that $\underset{\sim}{Q}$ is congruent to $\underset{\sim}{R}$ and $\underset{\sim}{R}$ is congruent to $\underset{\sim}{S}$; that is,

$$\alpha_{RQ} = \beta_{QR} \quad \text{and} \quad \alpha_{SR} = \beta_{RS} \quad .$$

Then by Theorem 18 (§4.3),

$$\alpha_{SQ} = \alpha_{SR}\alpha_{RQ} = \beta_{RS}\beta_{QR} = \beta_{QR}\beta_{RS} = \beta_{QS} \quad ,$$

which shows that $\underset{\sim}{Q}$ is congruent to $\underset{\sim}{S}$. □

If $\underset{\sim}{Q}$ and $\underset{\sim}{S}$ are particles in *COL* which are not congruent we can define a particle $\underset{\sim}{T} \in \hat{\underset{\sim}{S}}$ whose natural time scale is defined such that

$$\underset{TS}{f^+}(x) \stackrel{def}{=} (\beta_{QS}/\alpha_{SQ})^{\frac{1}{2}}x = \underset{TS}{f^-}(x) \quad .$$

Then

$$\alpha_{TQ} = (\beta_{QS}\alpha_{SQ})^{\frac{1}{2}} = \beta_{QT} \quad ,$$

from which we see that $\underset{\sim}{Q}$ and $\underset{\sim}{T}$ are congruent. Since the natural time scale of each particle is only determined to within an arbitrary affine transformation, we could further specify the time scales of particles in a particular collinear set, by choosing a given particle, say $\underset{\sim}{Q} \in COL$, and specifying that each other particle in *COL* is congruent to $\underset{\sim}{Q}$. By the preceding theorem, all particles in the collinear set are now congruent to each other. Since this theorem only applies to one collinear set of particles, we must be careful not to apply the theorem to two or more distinct collinear sets of particles, for this would assume the transitivity of congruence for non-collinear particles.

Given two congruent particles $\underset{\sim}{Q},\underset{\sim}{S} \in COL$ such that $\underset{\sim}{Q} \parallel \underset{\sim}{S}$, their record functions are of the form:

$$\left(f_{QS} \circ f_{SQ}\right)^*(x) = x + 2d_{QS}\ , \text{ and}$$

$$\left(f_{SQ} \circ f_{QS}\right)^*(x) = x + 2d_{SQ}\ ,$$

where the constants d_{QS} and d_{SQ} are called the *directed distance of* $\underset{\sim}{S}$ *relative to* $\underset{\sim}{Q}$, and the *directed distance of* $\underset{\sim}{Q}$ *relative to* $\underset{\sim}{S}$, respectively. The directed distances are defined in terms of the time scales of the particles, and so they are not invariant with respect to transformations of natural time scales.

THEOREM 53 (Additivity of Directed Distances)
Let $\underset{\sim}{Q},\underset{\sim}{S},\underset{\sim}{T}$ *be congruent particles in a collinear set. If* $\underset{\sim}{Q} \parallel \underset{\sim}{S} \parallel \underset{\sim}{T}$, *then*

(i) $$d_{QT} = d_{QS} + d_{ST}\ , \text{ and}$$

(ii) $$d_{SQ} = -\, d_{QS}\ .$$

REMARK. It is important to note that, in contrast to the analogous property for collinear rapidities, this property only applies if $\underset{\sim}{Q},\underset{\sim}{S},\underset{\sim}{T}$ are congruent, since the directed distances are defined in terms of the time scales of the particles.

This theorem is a consequence of Theorems 18 (§4.3) and 52 (§8.2).

PROOF. Part (ii) is a special case of part (i) with $\underset{\sim}{T} = \underset{\sim}{Q}$, so it is only necessary to prove part (i). Since $\underset{\sim}{Q}, \underset{\sim}{S}, \underset{\sim}{T}$ are congruent, there are constants $\gamma_{SQ}, \gamma_{TS}, \delta_{QS}, \delta_{ST}$ such that

$$f^+_{SQ}(x) = x + \gamma_{SQ} \quad , \quad f^+_{TS}(x) = x + \gamma_{TS} \quad \text{and}$$

$$f^-_{QS}(x) = x + \delta_{QS} \quad , \quad f^-_{ST}(x) = x + \delta_{ST} \quad .$$

By Theorem 18 (§4.3),

$$f^+_{TQ}(x) = x + \gamma_{TS} + \gamma_{SQ} \quad \text{and} \quad f^-_{QT}(x) = x + \delta_{QS} + \delta_{ST} \ .$$

Thus

$$\begin{aligned}\left(f_{QT} \circ f_{TQ}\right)^*(x) &= x + \left(\gamma_{SQ} + \delta_{QS}\right) + \left(\gamma_{TS} + \delta_{ST}\right) \\ &= x + 2d_{QS} + 2d_{ST} \ ,\end{aligned}$$

since $\gamma_{SQ} + \delta_{QS} = 2d_{QS}$ and $\gamma_{TS} + \delta_{ST} = 2d_{ST}$. □

§8.3 Partitioning a Collinear Set Into Synchronous Equivalence Classes.

Given any particles $\underset{\sim}{Q}, \underset{\sim}{S} \in COL$ we say that $\underset{\sim}{Q}$ and $\underset{\sim}{S}$ are *synchronous* if

$$f^+_{QS}(x) = f^-_{SQ}(x) \quad \text{and} \quad f^-_{QS}(x) = f^+_{SQ}(x)$$

(One condition implies the other, by Theorem 18 (§4.3)).

THEOREM 54 (Synchronous Parallel Particles)
The synchronous relation is an equivalence relation on any collinear class of parallels.

This theorem is a consequence of Theorem 18 (§4.3).

PROOF. By definition, the synchronous relation is reflexive and symmetric. In order to show that the synchronous relation is transitive, we consider three particles $\underset{\sim}{Q}, \underset{\sim}{S}, \underset{\sim}{T} \in \mathcal{COL}$ such that $\underset{\sim}{Q} \parallel \underset{\sim}{S} \parallel \underset{\sim}{T}$ and such that the pairs $\underset{\sim}{Q}, \underset{\sim}{S}$ and $\underset{\sim}{S}, \underset{\sim}{T}$ are synchronous. It then follows from the definition of directed distance, that

$$f^+_{QS}(x) = f^-_{SQ}(x) = x + d_{QS} \quad \text{and} \quad f^+_{ST}(x) = f^-_{TS}(x) = x + d_{ST} \; .$$

Then, by Theorem 18 (§4.3),

$$\begin{aligned} f^+_{QT}(x) = f^+_{QS} \circ f^+_{ST}(x) &= x + d_{QS} + d_{ST} \\ &= x + d_{ST} + d_{QS} \\ &= f^-_{TS} \circ f^-_{SQ}(x) \\ &= f^-_{TQ}(x) \; . \quad \square \end{aligned}$$

This result also follows from Theorem 43 (§7.3).

THEOREM 55 (Synchronous Collinear Sub-SPRAYs)
The synchronous relation is an equivalence relation on any collinear sub-SPRAY.

This theorem is a consequence of Theorem 18 (§4.3).

PROOF. By definition, the synchronous relation is reflexive and symmetric. To show that it is transitive on the set of particles belonging to a collinear sub-SPRAY, we consider three particles $\underset{\sim}{Q},\underset{\sim}{S},\underset{\sim}{T} \in CSP$ such that the pairs $\underset{\sim}{Q},\underset{\sim}{S}$ and $\underset{\sim}{S},\underset{\sim}{T}$ are synchronous. At the event of coincidence, the real index of $\underset{\sim}{Q}$ must be the same as the real index of $\underset{\sim}{S}$, which must also be the same as the real index of $\underset{\sim}{T}$, since both pairs are synchronous. Therefore the modified signal functions are of the form:

$$f^+_{QS}(x) = f^-_{SQ}(x) = \alpha(x-a) + a \text{ , and}$$

$$f^+_{ST}(x) = f^-_{TS}(x) = \beta(x-a) + a \text{ .}$$

By Theorem 18 (§4.3),

$$f^+_{QT}(x) = f^+_{QS} \circ f^+_{ST}(x) = \alpha\beta(x-a) + a = f^-_{TS} \circ f^-_{SQ}(x) = f^-_{TQ}(x) \quad . \qquad \square$$

We have shown that the synchronous relation is an equivalence relation on any class of parallels and on any collinear sub-SPRAY: these are the only subsets of a collinear set of particles on which the synchronous relation is an equivalence relation. Thus it is not possible for all the particles of a collinear set to be synchronous. However, all particles of a collinear sub-SPRAY could be synchronous, and each class of parallels could be synchronous with that member which is contained in the given synchronous collinear sub-SPRAY.

§8.4 Coordinate Frames in a Collinear Set

Let $\{\underset{\sim}{S}^{\alpha}$: *α real*, $\underset{\sim}{S}^{\alpha} \in$ *COL*} be an indexed class of parallels in *COL*. Given any particle $\underset{\sim}{S}^{x}$ and any instant $S^{x}_{t} \in \underset{\sim}{S}^{x}$, the coefficients of the ordered pair *(x;t)* of reals are called the *position-time coordinates* of the event $[S^{x}_{t}]$. The set of all events in *col*, indexed by the corresponding ordered pairs of reals, is called a *coordinate frame in col;* the event *(0,0)* is called the *origin in position-time* of the coordinate frame; and the set of events {*(0,t): t real*} is called the *origin in position* of the coordinate frame.

THEOREM 56 (Some Useful Kinematic Relations)

Let $\{\underset{\sim}{Q}^{\alpha}\colon \alpha \text{ real}, \underset{\sim}{Q}^{\alpha} \in COL\}$ and $\{\underset{\sim}{S}^{\beta}\colon \beta \text{ real}, \underset{\sim}{S}^{\beta} \in COL\}$ be two *distinct indexed classes of parallels in COL such that:*

(i) $\underset{\sim}{Q}^{0}$ *and* $\underset{\sim}{S}^{0}$ *are synchronous,*

(ii) $Q^{0}_{0} \simeq S^{0}_{0}$ *, and*

(iii) $\underset{\sim}{Q}^{0} \not\parallel \underset{\sim}{S}^{0}$.

Let r *be the directed rapidity of* $\{\underset{\sim}{Q}^{\alpha}\}$ *with respect to* $\{\underset{\sim}{S}^{\beta}\}$.

For any real x*, let* u, w, y *be real numbers such that*

$$f^{+}_{S^x S^0}(u) = t\ ,\quad S^{x}_{t} \simeq Q^{0}_{w},\quad \text{and}\quad f^{-}_{S^0 S^x}(t) = y$$

(see Figure 57).

Then

(i) $x/t = \tanh r \overset{def}{=} v$, *where* v *is the "velocity" of* $\{\underset{\sim}{Q}^{\alpha}\}$ *with respect to* $\{\underset{\sim}{S}^{\beta}\}$,

(ii) $w = e^{r}u = t \operatorname{sech} r$, *and*

(iii) $y = te^{r} \operatorname{sech} r$.

This theorem is a consequence of Theorem 50 (§8.1). It is used in the proof of Theorems 57 (§9.1), 58 (§9.3) and Corollary 1 to 58 (§9.3), and Theorems 59 (§9.4), 60 (§9.5) and 62 (§9.6).

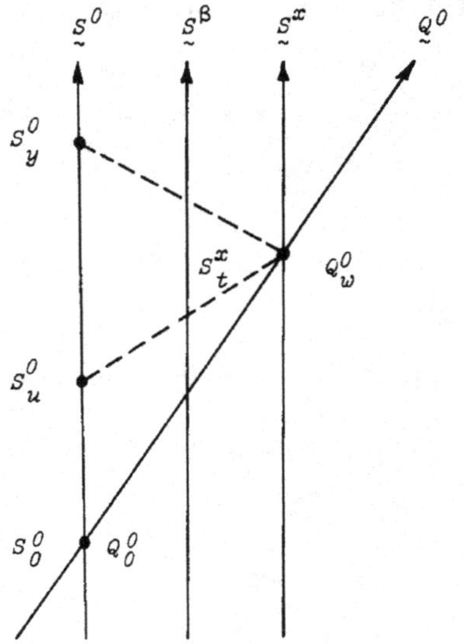

Fig. 57 This diagram illustrates the case where $x > 0$.

PROOF. By Theorem 50 (i) (§8.1) and the definition of directed rapidity (§8.1),

(1) $$\left[f_{Q^0 S^0} \circ f_{S^0 Q^0} \right]^* (u) = e^{2r} u$$

Since $\underset{\sim}{S}^0$ and $\underset{\sim}{S}^x$ are synchronous,

(2) $$t = f^+_{S^x S^0} (u) = u + x \text{ , and}$$

(3) $$y = f^-_{S^0 S^x} (t) = t + x = e^{2r} u \text{ .}$$

By equations (2) and (3),

(4) $$x/t = \tanh r = v \ .$$

Since $\underset{\sim}{Q}^0$ and $\underset{\sim}{S}^0$ are synchronous, equation (1) implies that

(5) $$w = f^{+}_{Q^0 S^0}(u) = e^r u \ , \text{ and}$$

(6) $$y = f^{-}_{S^0 Q^0}(w) = e^r w \ .$$

By equations (3), (4) and (5),

(7) $$w = e^r u = e^{-r} t(1+\tanh r) = t \operatorname{sech} r \ .$$

Equations (4) and (7) correspond to parts (i) and (ii), and part (iii) is obtained by combining equations (6) and (7). □

COROLLARY (Kinematics of Optical Lines)
The set of events in col which are on the right, or left, optical lines through the origin in position-time, have position-time coordinates which are related by the equations:

$$x/t = \lim_{r\to\infty}(\tanh r) = 1 \quad \text{and} \quad x/t = \lim_{r\to-\infty}(\tanh r) = -1 \ ,$$

respectively. Thus, signals have "infinite rapidity".

PROOF. By the Indexing Theorem (Theorem 41, §7.3), it follows that $x/t = \pm 1$, for right and left optical lines, respectively. That is, optical lines have unit velocity, which corresponds to infinite rapidity. □

CHAPTER 9

THREE-DIMENSIONAL KINEMATICS

Whereas the previous discussion had some similarity to the theory of absolute geometry, this final chapter departs radically from both absolute geometry and the more usual discussions of Minkowski space-time. Two ideas are of central importance; namely, that the velocity space of Minkowski space-time is hyperbolic, in contrast to the euclidean velocity space of Newtonian kinematics, and that space-time coordinates are related to homogeneous coordinates in a three-dimensional hyperbolic space.

It is shown that each SPRAY is a three-dimensional hyperbolic space with particles corresponding to "points" and with relative velocity as a metric function. Homogeneous coordinates in three-dimensional hyperbolic space correspond to space-time coordinates of particles in a SPRAY. This correspondence is eventually extended to all events and gives rise to the concept of a coordinate frame. The position space associated with each coordinate frame is shown to be a three-dimensional euclidean space, so the present axiomatic system is also an axiom system for euclidean geometry.

Transformations between homogeneous coordinate systems correspond to homogeneous Lorentz transformations, from which the inhomogeneous Lorentz transformations are derived. In conclusion we describe the trajectories of particles and optical lines relative to any coordinate frame.

§9.1 Each 3-SPRAY is a 3-Dimensional Hyperbolic Space

It has been known for some time that the velocity space of special relativity is hyperbolic; see, for example, the early references given by Pauli [1921, p.74]. The phenomena of spherical aberration and Thomas precession are simple consequences of the velocity space being hyperbolic and they have been discussed recently by Boyer [1965], Fock [1964] and Smorodinsky [1965].

In the next theorem, we show that each SPRAY is a metric space with observers being the "points" of the space and with relative rapidity as an intrinsic metric. We make the following definition: a *3-SPRAY* (denoted *3SP*[ ] is a SPRAY which has a maximal symmetric sub-SPRAY of four distinct particles. The existence of at least one 3-SPRAY is postulated in the Axiom of Dimension (Axiom VIII, §2.10). In a following theorem (Theorem 61, §9.5) we will show that each SPRAY is a 3-SPRAY.

THEOREM 57. *Each 3-SPRAY is a hyperbolic space of three dimensions with curvature of -1: the "points" of the space are the observers of the 3-SPRAY and relative rapidity is an intrinsic metric.*

This theorem is a consequence of Axioms IV (§2.4), VII (§2.9), XI (§2.13) and Theorem 22 (§5.2), the Corollary to Theorem 24 (§5.3), and Theorems 49 (§7.5), 51 (§8.1) and 56 (§8.4). It is used in the proof of Theorems 58 (§9.3) 60 (§9.4) and 63 (§9.7).

PROOF. A characterisation of 3-dimensional euclidean and hyperbolic spaces is given in Appendix 1. We will first show that all the conditions of this characterisation are satisfied by any 3-SPRAY, and then it will follow that each 3-SPRAY is either a 3-dimensional euclidean space or a 3-dimensional hyperbolic space. Definitions of concepts which have not yet been defined will be found in Appendix 1.

To show that a given SPRAY *(SPR)* is a metric space having relative rapidity as an intrinsic metric, we consider any three particles $\underset{\sim}{Q},\underset{\sim}{S},\underset{\sim}{T} \in SPR$. By the definition of relative rapidity (§8.1),

$$|r_{QS}| = |r_{SQ}|$$

and also

$$|r_{QS}| = 0 \text{ if and only in } \underset{\sim}{Q} \simeq \underset{\sim}{S}.$$

The triangle inequality for relative rapidity is a consequence of the Triangle Inequality (Axiom, IV, §2.4) which implies that

$$(1) \qquad f_{QT} \circ f_{TQ} \leqslant f_{QS} \circ f_{ST} \circ f_{TS} \circ f_{SQ} .$$

For any three particles $\underset{\sim}{Q}, \underset{\sim}{S}, \underset{\sim}{T} \in SPR$ there are instants $Q_a \in \underset{\sim}{Q}$, $S_b \in \underset{\sim}{S}$, $T_c \in \underset{\sim}{T}$ such that

$$Q_a \simeq S_b \simeq T_c$$

and by Theorem 49 (§7.5), the signal functions after coincidence have the form

$$f_{TQ}(Q_x) = T_y \text{ where } y = c + \alpha_{TQ}(x-a) ,$$

and so on. As in the proof of Theorem 51 (§8.1),

$$|r_{QT}| = \tfrac{1}{2}ln(\alpha_{QT} \cdot \alpha_{TQ}) ,$$

so by the inequality (1),

$$\begin{aligned} r_{QT} &= \tfrac{1}{2}ln(\alpha_{QT} \cdot \alpha_{TQ}) \\ &\leq \tfrac{1}{2}ln(\alpha_{QS} \cdot \alpha_{ST} \cdot \alpha_{TS} \cdot \alpha_{SQ}) \\ &= |r_{QS}| + |r_{ST}| . \end{aligned}$$

We have now shown that relative rapidity is a metric function and, by the Addition Law for Directed Rapidity (Theorem 51, §8.1), it follows that relative rapidity is an intrinsic metric.

Each SPRAY is unbounded (as a consequence of the Corollary to Theorem 24, §5.3), locally compact (by the Axiom of Compactness of Bounded sub-SPRAYs (Axiom XI, §2.13)), arcwise-connected (by Theorem 22, §5.2) and isotropic (by the Axiom of Isotropy of SPRAYs (Axiom VII, §2.9)). Now by the conclusion of Appendix 1, it follows that each 3-SPRAY is either a euclidean or a hyperbolic space of three dimensions.

We now consider a given *3-SPRAY (3SP)* and any particles $\underset{\sim}{Q}, \underset{\sim}{S} \in 3SP$. Homogeneous coordinates in three-dimensional euclidean and hyperbolic spaces are described in Appendix 2, to which we will refer. Let $(x;t)$ be position-time coordinates relative to $\underset{\sim}{S}$ in $col[\underset{\sim}{S}, \underset{\sim}{Q}]$ of an event coincident with $\underset{\sim}{Q}$, as in Theorem 56 (§8.4). Equation (i) of this theorem is

$$(2) \qquad x/t = \tanh r = v ,$$

which shows that t and x are homogeneous coordinates of $\underset{\sim}{Q}$ relative to $\underset{\sim}{S}$ in either a hyperbolic, or a euclidean, one-dimensional sub-space of $3SP$; with t corresponding to x_0 and x corresponding to x_1 of equations (1) or (3) of Appendix 2, according as to whether $3SP$ is euclidean or hyperbolic. We will now show that $3SP$ is hyperbolic by assuming the contrary and deducing a contradiction:
if $3SP$ is a euclidean space, a comparison of equation (2) with equation (3) of Appendix 2 shows that relative velocity is an intrinsic metric so, for collinear particles $\underset{\sim}{Q}, \underset{\sim}{S}, \underset{\sim}{T} \in 3SP$.

$$(3) \qquad v_{QT} = v_{QS} + v_{ST} ,$$

and by Theorem 51(§8.1),

$$(4) \qquad r_{QT} = r_{QS} + r_{ST} .$$

Now the general solution to Cauchy's functional equation

$$g(x+y) = g(x) + g(y)$$

(where g is a continuous function for positive real variables as discussed by Aczèl [1966]) is

$$g(x) = \alpha x \ , \qquad \alpha \ real \ .$$

Equation (2) implies that $v(r)$ is a continuous function and equations (3) and (4) show that $v(r)$ satisfies Cauchy's functional equation. Consequently there is a real constant α such that

$$v(r) = \alpha r \ ,$$

which is a contradiction of (2). Thus *3SP* can not be a euclidean space and so we conclude that *3SP* is a hyperbolic space.

Moreover, if we now compare equation (2) with equation (1) of Appendix 2, we see that each 3-SPRAY, equipped with relative rapidity as an (intrinsic) metric, is a hyperbolic space with curvature of -1. □

§9.2 Transformations of Homogeneous Coordinates in Three-Dimensional Hyperbolic Space

Transformations between sets of homogeneous coordinates in the 3-dimensional hyperbolic space, H_3, are of the form (see Appendix 2):

$$(1)\qquad \bar{x}_i = \sum_{k=0}^{3} a_{ik}\, x_k \qquad (i=0,1,2,3)\ ,$$

with $det[a_{ij}] \neq 0$.

Hyperbolic distance is independent of coordinate representations, so for any two points $x,y \in H_3$,

$$h(x,y) = h(\bar{x},\bar{y})\ ,$$

whence, by equation (2) of Appendix 2,

$$Arcosh\left\{|\Omega(x,y)|\left[\Omega(x,x)\Omega(y,y)\right]^{-\frac{1}{2}}\right\} = Arcosh\left\{|\Omega(\bar{x},\bar{y})|\left[\Omega(\bar{x},\bar{x})\Omega(\bar{y},\bar{y})\right]^{-\frac{1}{2}}\right\}$$

and so

$$\Omega^2(x,y)\Omega(\bar{x},\bar{x})\Omega(\bar{y},\bar{y}) = \Omega^2(\bar{x},\bar{y})\Omega(x,x)\Omega(y,y)\ .$$

Thus

$$\begin{aligned} &\left[-x_0y_0+x_\alpha y_\alpha\right]^2\left[-(a_{0j}x_j)^2 + (a_{1j}x_j)^2 + (a_{2j}x_j)^2 + (a_{3j}x_j)^2\right] \times \\ &\qquad \times \left[-(a_{0k}y_k)^2 + (a_{1k}y_k)^2 + (a_{2k}y_k)^2\right] \\ &= \left[-a_{0j}a_{0k}x_jy_k + a_{\beta j}a_{\beta k}x_jx_k\right]^2\left[-x_0^2+x_1^2+x_2^2+x_3^2\right]\left[-y_0^2+y_1^2+y_2^2+y_3^2\right]\ , \end{aligned}$$

where repeated indices imply a summation convention: Latin indices take the values $0,1,2,3$ and Greek indices take the values $1,2,3$. Equating coefficients of $x_0^4 y_0^4$, of $x_{(\beta)}^4 y_{(\beta)}^4$ (with no sum over β), and of $x_{(i)}^2 x_{(\beta)}^2 y_{(i)}^2 y_{(\beta)}^2$ with $\beta = 1,2,3$ and $i = 0,1,2,3$ and $\beta \neq i$ (and with no sum over β and i), we find that, since the argument of *Arcosh* must be real,

$$(2)\qquad \sum_{\alpha=1}^{3} a_{\alpha 0}^2 - a_{00}^2 = -1 ,$$

$$\sum_{\alpha=1}^{3} a_{\alpha\beta}^2 - a_{0\beta}^2 = 1 \qquad (\beta = 1,2,3) \text{ and}$$

$$\sum_{\alpha=1}^{3} a_{\alpha i} a_{\alpha k} - a_{0i} a_{0k} = 0 \qquad (i,k=0,1,2,3 \text{ and } i \neq k) ,$$

respectively. An immediate consequence of these equations is that $[a_{ij}]$ has an inverse $[\overset{*}{a}_{ij}]$ where

$$\overset{*}{a}_{00} = a_{00} , \quad \overset{*}{a}_{0\alpha} = -a_{\alpha 0} , \quad \overset{*}{a}_{\alpha 0} = -a_{0\alpha} , \quad \overset{*}{a}_{\alpha\beta} = a_{\beta\alpha} .$$

The matrix $[\overset{*}{a}_{ij}]$ represents the inverse coordinate transformation and the equations corresponding to equations (2) above are:

$$(3)\qquad \sum_{\alpha=1}^{3} a_{0\alpha}^2 - a_{00}^2 = -1 ,$$

$$\sum_{\alpha=1}^{3} a_{\beta\alpha}^2 - a_{\beta 0}^2 = 1 \qquad (\beta = 1,2,3) \text{ and}$$

$$\sum_{\alpha=1}^{3} a_{i\alpha} a_{k\alpha} - a_{i0} a_{k0} = 0 \qquad (i,k=0,1,2,3 \text{ and } i \neq k) .$$

Since $[a_{ij}]$ and $[\overset{*}{a}_{ij}]$ are inverses, it follows that

$$det[a_{ij}][\overset{*}{a}_{k\ell}] = (det[a_{ij}])^2 = 1 \ ,$$

whence

$$(4) \qquad det[a_{ij}] = det[\overset{*}{a}_{ij}] = \pm 1 \ .$$

The inverse of the transformation (1) is therefore given by

$$(5) \qquad x_i = \overset{*}{a}_{ij} \, \bar{x}_j$$

where

$$(6) \quad \overset{*}{a}_{00} = a_{00} \ , \ \overset{*}{a}_{0\alpha} = -a_{\alpha 0} \ , \ \overset{*}{a}_{\alpha 0} = -a_{0\alpha} \ , \ \overset{*}{a}_{\alpha\beta} = -a_{\beta\alpha} \ .$$

We can now verify that coordinate transformations represented by matrices having $a_{00} > 0$ form a group.

[Aside: In §9.6 below we will see that these transformations correspond to the *group of Lorentz transformations without time reversal*. In order to show that the transformations form a *group*, we observe first that the above matrices are non-singular so each matrix has a unique inverse. Secondly, if for any matrix $[a_{ij}]$, we define a matrix $[\bar{a}_{ij}]$ such that

$$\bar{a}_{00} \overset{def}{=} a_{00} \ , \ \bar{a}_{0\alpha} \overset{def}{=} -\sqrt{-1}\, a_{\alpha 0} \ , \ \bar{a}_{\alpha 0} \overset{def}{=} -\sqrt{-1}\, a_{0\alpha} \ , \ \bar{a}_{\alpha\beta} \overset{def}{=} -a_{\beta\alpha} \ ,$$

then equations (2) and (3) can be written in the equivalent forms

$$\bar{a}_{ij}\bar{a}_{kj} = \delta_{kj} \text{ and } \bar{a}_{ji}\bar{a}_{jk} = \delta_{ik} .$$

It is now easily verified that, if $[\bar{a}_{ij}]$ and $[\bar{b}_{ij}]$ are any two such matrices, then the product matrix $[\bar{c}_{ij}]$, where

$$\bar{c}_{ij} = \bar{a}_{ij}\bar{b}_{ij} ,$$

satisfies equations which are equivalent to equations (2), (3) and (4). It is now only necessary to show that the 00-terms of the original matrices are positive. Since

$$c_{ik} = a_{ij}b_{jk}$$

then

$$\begin{aligned} c_{00} &= a_{00}b_{00} + \sum_{\alpha} a_{0\alpha}b_{\alpha 0} \\ &\geqslant \left(1 + \sum_{\alpha} a_{0\alpha}^2\right)^{\frac{1}{2}}\left(1 + \sum_{\beta} b_{\beta 0}^2\right)^{\frac{1}{2}} - \left(\sum_{\alpha} a_{0\alpha}^2\right)^{\frac{1}{2}}\left(\sum_{\beta} b_{\beta 0}^2\right)^{\frac{1}{2}} \\ &> 0 \qquad , \end{aligned}$$

by the first of equations (2) and (3).]

A set of homogeneous coordinates has the point x as *origin* if x has the coordinates

$$x_i = (x_0, x_1, x_2, x_3) = (x_0, 0, 0, 0) .$$

Relative to the same set of coordinates, we denote the *coordinates of any point* $z \in H_3$ *with respect to a coordinate system having* x *as origin by*

$$z_i^x = (z_0^x, z_1^x, z_2^x, z_3^x)$$

and, in particular, the coordinates of the origin x are denoted by

$$x_i^x = (x_0^x, x_1^x, x_2^x, x_3^x) = (x_0^x, 0, 0, 0) \ .$$

A transformation of coordinates can therefore be expressed in the form:

(7) $$z_i^x = a_{ij}\, z_j^y \ , \quad z_i^y = a_{ij}^*\, z_j^x \ .$$

We now derive some results which will be used in following sections. A particular case of (7) is

(8) $$y_i^x = a_{ij}\, y_j^y = a_{io}\, y_0^y \ ,$$

so by equations (2),

(9) $$\left[(y_1^x)^2 + (y_2^x)^2 + (y_3^x)^2\right]^{\frac{1}{2}} = \left[a_{10}^2 + a_{20}^2 + a_{30}^2\right] |y_0^y| = \left[-1 + a_{00}^2\right]^{\frac{1}{2}} |y_0^y| \ .$$

Also, by equations (2) and (9),

(10) $$h(y,x) = Arcosh\left\{|\Omega(y_i^x, x_i^x)|\left[\Omega(y_i^x, y_i^x)\Omega(x_i^x, x_i^x)\right]^{-\frac{1}{2}}\right\} = Arcosh\ a_{00}$$

and so

(11) $$\sinh h(y,x) = (-1 + a_{00}^2)^{\frac{1}{2}} \quad .$$

Thus, from equations (9), (10), (11) we find that

(12) $$y_0^x = y_0^y \cosh h(y,x) \quad ,$$

(13) $$\left[(y_1^x)^2 + (y_2^x)^2 + (y_3^x)^2\right]^{\frac{1}{2}} = |y_0^y| \sinh h(y,x)$$

and

(14) $$\left[(y_1^x)^2 + (y_2^x)^2 + (y_3^x)^2\right]^{\frac{1}{2}} |y_0^x|^{-1} = \tanh h(y,x) \quad .$$

Coordinate transformations which have $a_{00} = 1$ leave the origin invariant, by equation (10); so we see that having specified an origin, say x, any choice of a system of homogeneous coordinates is arbitrary to within an equivalence class of transformations having

$$a_{00} = 1$$

and also, by equations (2) and (3),

$$a_{\alpha 0} = a_{0\alpha} = 0 \qquad (\alpha = 1,2,3) \quad ,$$

so equations (2) become

(15) $$\sum_{\beta=1}^{3} a_{\alpha\beta}\, a_{\gamma\beta} = \delta_{\alpha\gamma} \qquad (\alpha,\gamma = 1,2,3) ,$$

which are the conditions for orthogonal transformations in a three-dimensional euclidean space. The quadratic form

$$(z_1^x)^2 + (z_2^x)^2 + (z_3^x)^2$$

is therefore invariant with respect to the subgroup of coordinate transformations having $a_{00} = 1$.

In the following section, we will consider isometric mappings between each 3-SPRAY and three-dimensional hyperbolic space (H_3). To clarify the distinction between the 3-SPRAY and H_3, we will use corresponding upper and lower case symbols for corresponding particles and points. Thus, for example,

$$\underset{\sim}{Q}, \underset{\sim}{S}, \underset{\sim}{U}, \underset{\sim}{W} \in 3SP \text{ correspond to } q,s,u,w \in H_3$$

and, for any two particles $\underset{\sim}{Q}, \underset{\sim}{S} \in 3SP$,

$$|r_{QS}| = h(q,s) \quad .$$

§9.3 Space-Time Coordinates Within the Light Cone

In this section we will define space-time coordinates by establishing a correspondence between the *0*-component of homogeneous coordinates and the *t*-component of position-time coordinates which, we recall, apply only to the restricted case of "one-dimensional motion". The reader may already have noticed the similarity between the formulae of §9.2 and the Lorentz transformation formulae.

THEOREM 58. *Each 3-SPRAY is a hyperbolic space and so any particle in a given 3-SPRAY can be represented by a set of homogeneous coordinates. Now, given a particle* $\underset{\sim}{S}$ *with a natural time scale, we can define a mapping* τ_S *such that for each particle* $\underset{\sim}{Q} \in 3SP\,[S_0]$.

(1) $$\tau_S\colon\ q_0^s \to t = q_0^s$$

and then

(2) $$\left\{(q_1^s)^2 + (q_2^s)^2 + (q_3^s)^2\right\}^{\frac{1}{2}} = |x| \quad ,$$

where $(x;t)$ *are position-time coordinates, relative to* $\underset{\sim}{S}$ *in* $col[\underset{\sim}{Q},\underset{\sim}{S}]$, of any event $[Q_w]$ coincident with $\underset{\sim}{Q}$.

The quadruple $[q_0^s, q_1^s, q_2^s, q_3^s]$ is called a set of *space-time coordinates* of the event $[Q_w]$.

relative to a coordinate system having $\underset{\sim}{S}$ as an *origin in space* and $[S_0]$ as an *origin in space-time.* The 0-coordinate is called the *time coordinate* and the remaining three coordinates are called *space coordinates.* Thus *the space coordinates are determined to within an arbitrary orthogonal space transformation.*

Furthermore, for any two particles $\underset{\sim}{Q}, \underset{\sim}{U} \in 3SP[S_0]$ *such that* $[\underset{\sim}{Q}, \underset{\sim}{S}, \underset{\sim}{U}]$, *any two sets of space-time coordinates of events coincident with* $\underset{\sim}{Q}$ *and* $\underset{\sim}{U}$, *respectively, relative to the same coordinate system having* $\underset{\sim}{S}$ *as an origin in space and* $[S_0]$ *as an origin in space-time, are related by the set of equations*

$$(3) \qquad q_1^s : u_1^s = q_2^s : u_2^s = q_3^s : u_3^s \ .$$

In particular, if $[q_0^s, q_1^s, q_2^s, q_3^s]$ *and* $[\bar{q}_0^s, \bar{q}_1^s, \bar{q}_2^s, \bar{q}_3^s]$ *are the space-time coordinates, relative to the same coordinate system, of any two events coincident with* $\underset{\sim}{Q}$, *then*

$$(4) \qquad q_0^s : \bar{q}_0^s = q_1^s : \bar{q}_1^s = q_2^s : \bar{q}_2^s = q_3^s : \bar{q}_3^s$$

REMARK. An origin in space is only determined to within an equivalence class of permanently coincident synchronous particles.

This theorem is a consequence of Theorems 56 (§8.4) and 57 (§9.1). It is used in the proof of Theorems 59 (§9.4), 61 (§9.5) and 62 (§9.6).

PROOF. Let $\underset{\sim}{Q}$ be any particle in $3SP[S_0]$ and let

$$\{\underset{\sim}{S}^{\alpha} : \alpha \text{ real}, \underset{\sim}{S}^{\alpha} \in COL[\underset{\sim}{Q},\underset{\sim}{S}]\}$$

be an indexed class of parallels such that for all real t,

$$S_t^0 \simeq S_t \quad .$$

Now, by Theorem 56 (§8.4),

$$\{[Q_w] : Q_w \in \underset{\sim}{Q}\} = \{[S_t^x] : x/t = \tanh r = v\}$$

and we note in particular, that

(5) $$x/t = \tanh r = v \quad ,$$

where r is the directed rapidity of $\underset{\sim}{Q}$ and $\underset{\sim}{S}$. It is important to realise at this stage that $\underset{\sim}{Q}$ and $\underset{\sim}{S}$ are not necessarily synchronous; furthermore, equation (i) of Theorem 56 (§8.4) and the above equation (5) make statements which are independent of the time scale of $\underset{\sim}{Q}$.

Equation (14) of the preceding section can be written as

(6) $$\left[(q_1^s)^2 + (q_2^s)^2 + (q_3^s)^2\right]^{\frac{1}{2}} |q_0^s|^{-1} = \tanh h(q,s) \quad .$$

There is an obvious analogy between equations (5) and (6) which leads us to make the *coordinate time identification mapping*,

(7) $$\tau_{S,Q} : q_0^s \mapsto t = q_0^s \quad .$$

Since

$$|r_{QS}| = h(q,s) ,$$

equations (5) and (6) imply that

$$\left[(q_1^s)^2 + (q_2^s)^2 + (q_3^s)^2\right]^{\frac{1}{2}} = |x| , \qquad (8)$$

which is equivalent to equation (2). The coordinates q_1^s, q_2^s, q_3^s are thus determined only to within a class of coordinate transformations which leave (8) invariant. By equation (15) of the previous section, this is the class of orthogonal transformations in a three-dimensional euclidean space, which will be identified later in Theorem 60 (§9.4). Similar considerations apply for any particle $\underset{\sim}{Q} \in 3SP[S_0]$, so we define a mapping τ_S which is an obvious extension of the mapping defined by (7).

The collinearity conditions (3) are equivalent to the collinearity condition stated in Appendix 2. The final set of equations (4) is a consequence of representing points in three-dimensional hyperbolic space by classes of quadruples (see Appendix 2). □

COROLLARY 1. *If, furthermore,* $\underset{\sim}{Q}$ *and* $\underset{\sim}{S}$ *are synchronous and if, in* $col[\underset{\sim}{Q},\underset{\sim}{S}]$,

$$Q_w^0 \simeq S_t^x \quad ,$$

then the mapping τ_S *implies that*

$$w = q_0^q \quad ,$$

where q_0^q *is the 0-component of the set of coordinates* q_i^q .

This corollary is a consequence of Theorem 56 (§8.4). It is used in the proof of Theorems 59 (§9.4) and 62 (§9.6).

PROOF. By Theorem 56 (ii) of §8.4,

$$w = t \text{ sech } r$$

and this is analogous to equation (13) of the preceding section which is

$$q_0^q = q_0^s \text{ sech } h(q,s) \quad ,$$

whence

$$w = q_0^q \quad . \quad \square$$

If we let $[y_0, y_1, y_2, y_3]$ be the space-time coordinates of an event relative to a coordinate system having $\underset{\sim}{S}$ as an origin in space and $[S_0]$ as the origin in space-time, we see that the previous theorem applies only to events whose space-time coordinates satisfy the condition

$$y_1^2 + y_2^2 + y_3^2 < y_0^2 \quad ;$$

we say that these events are *within the light cone* whose *vertex* is $[S_0]$. Events within the light cone which have

$$y_0 > 0 \qquad (\text{or } y_0 < 0)$$

are said to be within the *upper* (or *lower*) *light cone* whose *vertex* is $[S_0]$.

COROLLARY 2 (Position Space)

Given a coordinate system which has $\underset{\sim}{S}$ *as an origin in space, a set of events represented by*

(1) $$\left\{[t,y_1,y_2,y_3] : t^2>y_1^2+y_2^2+y_3^2 : y_1,y_2,y_3 \text{ constant; } t \text{ real}\right\}$$

is the set of events, within the light cone, which coincide with some particle which is parallel to $\underset{\sim}{S}$*; and conversely, given a particle which is parallel to* $\underset{\sim}{S}$*, there are real numbers* y_1,y_2,y_3 *such that the set of all events, within the light cone, which coincide with this particle can be represented by (1).*

Thus any set of events (1) is the set of events, within the light cone, of an observer which is parallel to $\underset{\sim}{S}$*. We shall represent this set of events and the corresponding observer by the triple*

$$(y_1,y_2,y_3) \; .$$

For any given coordinate system, the corresponding *position-space* is defined to be the set of all particles which are parallel to, and synchronous with, the origin in space. *Thus position-space can be represented as*

$$\{(y_1,y_2,y_3) : y_1,y_2,y_3 \text{ real}\} \; .$$

REMARK. Each triple represents an equivalence class of permanently coincident particles which are parallel to, and synchronous with, the origin in space.

This corollary is a consequence of Corollary 1 of Theorem 33 (§6.4). It is used in the proof of Corollary 3 of Theorem 61 (§9.5) and Theorem 62 (§9.6).

PROOF. By equations (3) of the above theorem, the set of events (1) is contained in some col and by equation (2), all of these events are at the same distance from $\underset{\sim}{S}$. Since the given 3-SPRAY is a hyperbolic space, if we take any real number a with

$$a^2 - y_1^2 - y_2^2 - y_3^2 > 0 \quad ,$$

there is some particle $\underset{\sim}{Q} \in 3SP$ such that $\underset{\sim}{Q}$ coincides with the event whose coordinates are $[a, y_1, y_2, y_3]$. Now by equations (3) of the above theorem

$$col = col[\underset{\sim}{Q}, \underset{\sim}{S}]$$

and by equations (4) of the above theorem and Corollary 1 to Theorem 33 (§6.4), we see that the set of events (1) is on the same side of $\underset{\sim}{S}$ in $col[\underset{\sim}{Q}, \underset{\sim}{S}]$ and is therefore the set of events, within the light cone, which coincides with some parallel to $\underset{\sim}{S}$

Conversely, given any particle $\underset{\sim}{U}$ in position space, $\underset{\sim}{U}$ is parallel to $\underset{\sim}{S}$ and both $\underset{\sim}{U}$ and $\underset{\sim}{S}$ are contained in some *COL*. If $[y_0(w), y_1(w), y_2(w), y_3(w)]$ are the coordinates of any event $[U_w]$, equations (2) and (3) of the above theorem imply that

$$y_1(w) : y_1(0) = y_2(w) : y_2(0) = y_3(w) : y_3(0) = \pm 1 : 1 \quad .$$

Since $\underset{\sim}{U}$ can not cross $\underset{\sim}{S}$ in *col*, the positive sign must apply for all w, which completes the proof. □

§9.4 Properties of Position Space

Before establishing the main result of this section, we prove the following important property of the synchronous relation. This property is also applied in Theorem 63 (§9.7) where we show that each coordinate frame "can be calibrated in the same physical units".

THEOREM 59. *The synchronous relation is an equivalence relation on the set of particles of any SPRAY.*

This theorem is a consequence of Theorems 56 (§8.4), 58 (§9.3) and Corollary 1 of Theorem 58 (§9.3). It is used in the proof of Theorems 60 (§9.4) and 63 (§9.7).

PROOF. Clearly the synchronous relation is reflexive and symmetric, so we only have to prove transitivity. Let $\underset{\sim}{S}$ be any particle with a natural time scale and let $\underset{\sim}{Q}, \underset{\sim}{U} \in SPR[S_0]$

be any two particles which are synchronous with $\underset{\sim}{S}$, though not necessarily synchronous with each other. Then the homogeneous coordinates of any particle $\underset{\sim}{W} \in SPR[S_0]$, with respect to coordinate systems having $\underset{\sim}{S}, \underset{\sim}{Q}$ and $\underset{\sim}{U}$ as origins, are related by transformations having the same form as equations (7) of §9.2:

$$w_i^s = a_{ij}\, w_j^q \quad and \quad w_i^q = \overset{*}{a}_{ij}\, w_j^s \quad and$$

$$w_i^s = b_{ij}\, w_j^u \quad and \quad w_i^u = \overset{*}{b}_{ij}\, w_j^s \quad .$$

Thus, in particular,

$$q_i^u = \overset{*}{b}_{ij}\, a_{jk}\, q_k^q = \overset{*}{b}_{ij}\, a_{j0}\, q_0^q \quad ,$$

and similarly,

$$u_i^q = \overset{*}{a}_{ij}\, b_{jk}\, u_k^u = \overset{*}{a}_{ij}\, b_{j0}\, u_0^u$$

whence

$$q_0^u : q_0^q = u_0^q : u_0^u = a_{00}\, b_{00} - a_{\beta 0}\, b_{\beta 0} \overset{def}{=} \alpha \quad ,$$

where α is a constant. Now the pairs $\underset{\sim}{Q}, \underset{\sim}{S}$ and $\underset{\sim}{U}, \underset{\sim}{S}$ are synchronous so, as in Theorem 58 (§9.3) and Corollary 1 to Theorem 58 (§9.3), all terms in the above equation can be identified. The second of equations (ii) of Theorem 56 (§8.4) implies that a signal leaving $\underset{\sim}{Q}$ at $[Q_u]$ arrives at $\underset{\sim}{U}$ at $[U_w]$ where

$$w = u_0^u = \alpha^{-1} u_0^q = \alpha^{-1} t = u\alpha^{-1} e^r \cosh r$$

and similarly, a signal leaving $\underset{\sim}{U}$ at $[U_u]$ arrives at $\underset{\sim}{Q}$ at $[Q_w]$, where

$$w = q_0^q = \alpha^{-1} q_0^u = \alpha^{-1} t = u\alpha^{-1} e^r \cosh r \quad .$$

It follows that Q and U are synchronous. (It should be noted that the subscripts u and w have the same meanings as in the statement of Theorem 56 (§8.4): they do not refer to the particles $\underset{\sim}{U}$ or $\underset{\sim}{W}$ or to sets of homogeneous coordinates.) We have now shown that the synchronous relation is transitive. □

THEOREM 60 (Properties of Position Space)

(i) *Every position space is a three-dimensional euclidean space. If a position space is represented as*

$$\{(y_1, y_2, y_3) : y_1, y_2, y_3 \text{ real}\} \quad ,$$

then y_1, y_2, y_3 *are orthogonal cartesian coordinates.*

(ii) *Every position space is an equivalence class of parallel synchronous particles.*

This theorem is a consequence of Theorem 1 (§2.5), Corollary 2 of Theorem 33 (§6.4) and Theorems 46 (§7.5), 56(§8.4) and 57 (§9.1). It is used in the proof of Theorem 61 (§9.5).

PROOF. Let there by a y-coordinate system which has the particle $\underset{\sim}{S}$ as an origin in space and the event $[S_0]$ as an origin in space-time. Let $\underset{\sim}{Q}$ and $\underset{\sim}{U}$ be any two particles in $3SP[S_0]$ which are synchronous with $\underset{\sim}{S}$, and let

$$[1,0,0,0] \ , \quad [q_0,q_1,q_2,q_3] \ , \quad [u_0,u_1,u_2,u_3]$$

be the space-time coordinates in the y-coordinate system of events $[S_1]$, $[Q_w]$, $[U_z]$ such that a signal goes from $[S_1]$ to $[Q_w]$ and a signal goes from $[Q_w]$ to $[U_z]$. By the previous theorem and equation (i) of Theorem 56 (§ 8.4),

$$w = 1.\, exp\ |r_{QS}|$$

$$z = w.\, exp|\, r_{UQ}| = exp\left(|r_{UQ}| + |r_{QS}|\right)$$

and by equation (ii) of the same theorem,

$$q_0 = w\ cosh\ r_{QS} = exp\,|r_{QS}|\ cosh\ r_{QS} \quad ,$$

$$u_0 = z\ cosh\ r_{US} = exp\left(|r_{UQ}| + |r_{QS}|\right)\ cosh\ r_{US} \quad ,$$

whence

(1) $\quad u_0/q_0 = exp\,|r_{UQ}|\ cosh\ r_{US}\ sech\ r_{QS} \quad .$

By Theorem 57 (§9.1) and equation (2) of Appendix 2,

$$cosh\ r_{UQ} = |\Omega(u,q)|\,[\Omega(u,u)\Omega(q,q)]^{-\frac{1}{2}}$$

and so

$$exp\,|r_{UQ}| = [\Omega(u,u)\Omega(q,q)]^{\frac{1}{2}}[\,|\Omega(u,q)|+\{\Omega^2(u,q)-\Omega(q,q)\Omega(u,u)\}^{\frac{1}{2}}]\ ,$$

$$cosh\ r_{QS} = q_0[q_0^2-q_1^2-q_2^2-q_3^2]^{-\frac{1}{2}} = |\Omega(q,s)|[\Omega(q,q)\Omega(s,s)]^{-\frac{1}{2}}$$

and

$$cosh\ r_{US} = u_0[u_0^2-u_1^2-u_2^2-u_3^2]^{-\frac{1}{2}} = |\Omega(u,s)|[\Omega(u,u)\Omega(s,s)]^{-\frac{1}{2}}\ .$$

Substituting these relations in (1), simplifying and cross multiplying, we obtain

$$\Omega(u,u)-|\Omega(u,q)| = [\Omega^2(u,q)-\Omega(q,q)\Omega(u,u)]^{\frac{1}{2}}\ ,$$

which becomes after squaring, simplifying and dividing both sides by $u_0^2-u_1^2-u_2^2-u_3^2$ (which is not zero since $[U_z]$ is within the light cone whose vertex is $[S_0]$),

$$(u_0-q_0)^2 = (u_1-q_1)^2 + (u_2-q_2)^2 + (u_3-q_3)^2\ ,$$

whence

(3a) $$u_0 = q_0 + [(u_1-q_1)^2+(u_2-q_2)^2+(u_3-q_3)^2]^{\frac{1}{2}}\ ,$$

the positive square root being taken, since

$$[U_z] \geqslant [Q_w]$$

and, by Theorem 56 (§8.4) and Theorem 1 (§2.5), it follows that $u_0 \geqslant q_0$. The ambiguity of sign was introduced by the operation of squaring: the case with the negative square root,

(3b) $$u_0 = q_0 - [(u_1-q_1)^2+(u_2-q_2)^2+(u_3-q_3)^2]^{\frac{1}{2}} ,$$

corresponds to a signal which goes from $[S_1]$ to $[U_z]$ and from $[U_z]$ to $[Q_w]$.

It follows from equations (3a) and (3b) that any set of particles represented by

(4) $$\{(y_1,y_2,y_3) : y_i = \alpha q_i + (1-\alpha)u_i \; ; \; i=1,2,3 \; ; \; \alpha \text{ real}\}$$

is collinear before $[S_0]$ and after $[S_0]$. By Corollary 2 to Theorem 33 (§6.4), this set of particles is contained in some collinear set and, since no two distinct members coincide at any event, they are parallel by Theorem 46 (§7.5). Also, by equations (3a) and (3b), it follows that any two (parallel) particles in position-space are synchronous: this completes the proof of proposition (ii).

Since position space is an equivalence class of synchronous parallel particles, the distance between (q_1,q_2,q_3) and (u_1,u_2,u_3) is given by $|q_0 - u_0|$. It follows that equations (3a) and (3b) are forms of Pythagoras' Theorem in a three-dimensional euclidean space with orthogonal cartesian coordinates, which establishes proposition(i). □

COROLLARY (Orthogonal Transformations in Position Space)

Given a fixed origin in position space, all transformations between orthogonal cartesian coordinate systems are of the form

$$y_\alpha = a_{\alpha\beta}\, x_\beta \,, \quad x_\alpha = a_{\beta\alpha}\, y_\beta$$

where $[a_{\alpha\beta}]$ *is any orthogonal* 3×3*-matrix.*

This corollary is used in the proof of Corollary 3 of Theorem 61 (§9.5).

PROOF. By equations (7) and (15) of §9.2. The fixed natural time scale of the origin excludes transformations of the form

$$y_\alpha = \mu a_{\alpha\beta}\, x_\beta \,, \quad x_\alpha = \mu^{-1} a_{\beta\alpha}\, y_\beta \quad ,$$

where $\mu \neq 1$. □

As a consequence of the previous theorem, we can now define an equivalence relation of parallelism between position spaces. We say that two (or more) position spaces are *parallel* if all their particles are parallel, which means that their relative velocity is zero.

§9.5 Existence of Coordinate Frames

In the following theorem we show that space-time coordinates can be assigned to all events.

THEOREM 61 (Existence of Coordinate Frame)

Given a particle $\underset{\sim}{S}$ *with a natural time-scale, we can define space-time coordinates for all events, relative to a coordinate system having* $\underset{\sim}{S}$ *as an origin in space and* $[S_0]$ *as an origin in space-time. Conversely, any ordered quadruple* $[y_0, y_1, y_2, y_3]$ *is the set of coordinates of some event relative to this coordinate system.*

Any such correspondence between ordered quadruples of real numbers and events will be called a *coordinate frame.*

Furthermore, given any particle $\underset{\sim}{T}$ *which coincides with* $\underset{\sim}{S}$ *at some event, there are constants* a, v_0, v_1, v_2, v_3 *such that the set of events coincident with the particle* $\underset{\sim}{T}$ *can be represented by*

(1) $$\{[\lambda v_0 - a, \lambda v_1, \lambda v_2, \lambda v_3] : -\infty < \lambda < +\infty\}$$

with respect to the given coordinate frame.

Two events with coordinates $[y_0, y_1, y_2, y_3]$ *and* $[y_1', y_2', y_3', y_4']$ *are signal-related if and only if*

(2) $$(y_0' - y_0)^2 = (y_1' - y_1)^2 + (y_2' - y_2)^2 + (y_3' - y_3)^2 \quad .$$

This theorem is a consequence of Axioms I (§2.2), VIII (§2.10) and X (§2.12), Corollary 2 of Theorem 33 (§6.4) and Theorems 36 (§7.1), 48 (§7.5), 58 (§9.3) and 60 (§9.4). It is used in the proof of Theorems 62 (§9.6), 63 (§9.7), 64 (§9.7) and 65 (§9.7).

PROOF. (i) We begin by considering the case of a particle $\underset{\sim}{S}$ which has some instant $S_0 \in \underset{\sim}{S}$ such that $SPR[S_0]$ is a 3-SPRAY. (In part (ii) of this proof, we will show that all particles have this property). By Theorem 58 (§9.3), there is a coordinate system which has $\underset{\sim}{S}$ as an origin in space and $[S_0]$ as an origin in space-time.

We first show that any given event, say $[U_c]$, coincides with some particle in the position-space whose origin is $\hat{\underset{\sim}{S}}$. The case of an event which coincides with the origin in space is trivial, so we consider the case of an event which does not coincide with the origin in space. By Corollary 2 of Theorem 33 (§6.4), such an event and the origin in space are contained in a unique collinear set. Theorem 36 (§7.1) implies the existence of a particle $\underset{\sim}{V}$ which coincides with the given event and is parallel to the origin in space. Now take a particle $\underset{\sim}{W} \in \hat{\underset{\sim}{V}}$ such that $\underset{\sim}{W}$ is synchronous with $\underset{\sim}{S}$.

In position space, the particle $\underset{\sim}{W}$ has coordinates, say (y_1, y_2, y_3) . The Signal Axiom (Axiom I, §2.2) implies that, coincident with the origin in space, there is an event $[S_b]$

with coordinates $[x_0, 0, 0, 0]$ such that $[S_b]\ \sigma\ [U_c]$. We now define the space-time coordinates of $[U_c]$ to be :

$$[x_0 + (y_1^2+y_2^2+y_3^2)^{\frac{1}{2}}\ ,\ y_1\ ,\ y_2\ ,\ y_3\ ]$$

and we observe that this definition corresponds with the previous definition of §9.3, within the light cone, by equation (3a) of the proof of the previous theorem.

Conversely, given an ordered quadruple $[y_0, y_1, y_2, y_3]$, there is a particle $\underset{\sim}{W}$ in position space with coordinates (y_1, y_2, y_3) and, by the Signal Axiom (Axiom I §2.2), there is an instant $W_c \in \underset{\sim}{W}$ and some event $[S_b]$ which is coincident with the origin in space and which has coordinates $[y_0 - (y_1^2+y_2^2+y_3^2)^{\frac{1}{2}}\ ,\ 0, 0, 0]$, such that

$$[S_b]\ \sigma\ [W_c]\ .$$

Thus $[W_c]$ is an event which has coordinates $[y_0, y_1, y_2, y_3]$.

It is worth noting that the *1, 2, 3*-components of each event are the coordinates of a particle in position space which coincides with the event; the *0*-component of the event is equal to the numerical index of the corresponding instant from the particle in position space. Thus equations (3a) and (3b) of the proof of Theorem 60 (§9.4) apply to all signal-related pairs of events, which establishes equations (2) in this case (i).

In order to establish (1) for this case (i) we will first show that, for any instant $S_a \in \underset{\sim}{S}$, $SPR[S_a]$ is a 3-SPRAY. Theorem 36 (§7.1) implies that, for any particle $\underset{\sim}{Q} \in SPR[S_a]$, there is a particle $\underset{\sim}{R} \in 3SP[S_0]$ such that $\underset{\sim}{R} \parallel \underset{\sim}{Q}$ and conversely, given any particle $\underset{\sim}{R} \in 3SP[S_0]$ there is a particle $\underset{\sim}{Q} \in SPR[S_a]$ such that $\underset{\sim}{Q} \parallel \underset{\sim}{R}$. Equations (4) of Theorem 58 (§9.3) imply that there are constants v_0, v_1, v_2, v_3 such that the set of events coincident with the particle $\underset{\sim}{R}$ can be represented by

$$(3) \qquad \{[\lambda v_0, \lambda v_1, \lambda v_2, \lambda v_3] : -\infty < \lambda < \infty\}$$

By Theorem 48 (§7.5), there is a time displacement mapping

$$\tau : col[\underset{\sim}{Q}, \underset{\sim}{S}] \to col[\underset{\sim}{Q}, \underset{\sim}{S}]$$

such that

$$\tau(\underset{\sim}{R}) \simeq \underset{\sim}{Q} \ ;$$

furthermore, τ translates the 0-component and leaves the sum of the squares of the $1, 2, 3$-components invariant. Since $\underset{\sim}{Q}$ and $\underset{\sim}{R}$ are both contained in $col[\underset{\sim}{Q}, \underset{\sim}{S}]$, their $1, 2, 3$-components are proportional, by equations (3) of Theorem 58 (§9.3). Also $\underset{\sim}{Q} \parallel \underset{\sim}{R}$, so the previous two conditions imply that the $1, 2, 3$ components are invariant with respect to the mapping τ, since otherwise $\underset{\sim}{Q}$ and $\underset{\sim}{R}$ would cross at some event. Therefore, for all real λ,

(4) $\tau : [\lambda v_0, \lambda v_1, \lambda v_2, \lambda v_3] \rightarrow [\lambda v_0 - a, \lambda v_1, \lambda v_2, \lambda v_3]$.

Thus, for particles contained in $SPR[S_a]$, the coordinates

(5) $x_0' = x_0 - a$, $x_1' = x_1$, $x_2' = x_2$, $x_3' = x_3$

are homogeneous coordinates, and the mapping τ can be extended to a bijection between $3SP[S_0]$ and $SPR[S_a]$. Therefore $SPR[S_a]$ is a 3-SPRAY.

(ii) We will now show that each SPRAY is a 3-SPRAY by showing that every particle has the property assumed in (i). There is at least one 3-SPRAY as postulated in the Axiom of Dimension (Axiom VIII, §2.10). Now the Axiom of Connectedness (Axiom X, §2.12) implies that any event can be "connected to" this 3-SPRAY by two particles. The result (ii) above, applied twice, implies that the SPRAY specified by the given event is a 3-SPRAY.

(iii) Thus each SPRAY satisfies the assumption made in (i). The mapping (4) implies (1). □

We immediately have the following:

COROLLARY 1. *Each SPRAY is a 3-SPRAY.* □

Consequently, *the results of the previous theorems apply to any SPRAY.*

COROLLARY 2 (Time Coordinate Transformation)

Given any coordinate frame and any real number a, *there is a coordinate frame which is related to the given coordinate frame by the coordinate transformation*

$$\tau : [z_0, z_1, z_2, z_3] \to [z_0 - a, z_1, z_2, z_3] \quad ,$$

PROOF. This is a consequence of equations (5) of the above theorem. □

COROLLARY 3 (Coordinate Transformations in Position Space)

Given any coordinate frame, any quadruple of real numbers $[b_0, b_1, b_2, b_3]$ *and any orthogonal 3×3-matrix* $[a_{\alpha\beta}]$, *there is a coordinate frame, having the same euclidean metric defined on a parallel position space, which is related to the given coordinate frame by the coordinate transformations*

$$(1) \qquad \Delta : z_0 \to z_0 + b_0 \; , \quad z_\alpha \to b_\alpha + a_{\alpha\beta}\, z_\beta \; .$$

Conversely, given any two coordinate frames with synchronous parallel position spaces (that is, parallel position spaces having the same euclidean metric), the two coordinate frames are related by the transformations (1), where $[b_0, b_1, b_2, b_3]$ *is some quadruple of real numbers and* $[a_{\alpha\beta}]$ *is some orthogonal 3×3-matrix.*

This corollary is a consequence of Corollary 2 of Theorem 58 (§9.3) and the Corollary of Theorem 60 (§9.4). It is used in the proof of Theorems 62 (§9.6) and 63 (§9.7).

PROOF. By Corollary 2 to Theorem 58 (§9.3), there is a particle in position space with coordinates $(-b_1, -b_2, -b_3)$. By the above theorem, there is a coordinate frame which has this particle as an origin in space and the event, whose coordinates are $[-b_0, -b_1, -b_2, -b_3]$ in the given coordinate frame, as an origin in space-time. The two coordinate frames have parallel position spaces.

By the corollary to the previous theorem, the set of all isometric transformations between (orthogonal cartesian) coordinate systems in position spaces, having these two particles as origins, is the set of all space coordinate transformations of the form

(2) $$\Delta : z_\alpha \to b_\alpha + a_{\alpha\beta} z_\beta \qquad (\alpha = 1, 2, 3)$$

where $[a_{\alpha\beta}]$ is any orthogonal 3×3-matrix.

The only space-time coordinate transformations which are consistent with the transformations (2) and with equations (2) of the above theorem have

(3) $$\Delta : z_0 \to c + z_0 \quad ,$$

where c is a constant. Clearly $c = b_0$.

The proof of the converse proposition is similar. □

COROLLARY 4. *Given a coordinate frame and a positive real number* μ, *there is a coordinate frame which is related to the given coordinate frame by the transformations*

(1) $$x_i = \mu\, w_i \quad .$$

This corollary is used in the proof of Theorem 63 (§9.7).

PROOF. We shall call the given coordinate frame the w-coordinate frame. There is some particle $\underset{\sim}{Q}$, permanently coincident with the origin in space of the w-coordinate frame, which has a natural time-scale such that the coordinates of an event $[Q_\lambda]$ are $[\mu\lambda, 0, 0, 0]$. By the above theorem there is a coordinate frame, which we shall call the y-coordinate frame, which has $\underset{\sim}{Q}$ as an origin in space. Equations (2) of the above theorem apply to the y-coordinate frame as well as to the x-coordinate frame, so

(2) $$y_1^2 + y_2^2 + y_3^2 = \mu^2 (w_1^2 + w_2^2 + w_3^2) \quad .$$

Therefore there is some orthogonal 3×3 matrix $[a_{\alpha\beta}]$ such that

(3) $$y_\alpha = \mu\, a_{\alpha\beta}\, w_\beta \quad .$$

In accordance with the previous corollary, we define an x-coordinate frame such that

(4) $$x_0 = y_0 \quad \text{and} \quad x_\alpha = a_{\beta\alpha}\, y_\beta \quad .$$

Combining equations (3) and (4) shows that the x-coordinate frame is related to the y-coordinate frame by the transformations (1). □

§9.6 Homogeneous Transformations of Space-Time Coordinates

Having established the relationships between space-time coordinates and homogeneous coordinates of particles in each (three-dimensional hyperbolic) SPRAY, the homogeneous Lorentz transformation formulae can be derived by considering transformations of homogeneous coordinate systems in three-dimensional hyperbolic space.

THEOREM 62 (Homogeneous Lorentz Transformations)

Let $\underset{\sim}{Q}$ *and* $\underset{\sim}{S}$ *be two distinct synchronous particles with instants* $Q_0 \in \underset{\sim}{Q}$ *and* $S_0 \in \underset{\sim}{S}$ *such that* $Q_0 \simeq S_0$. *Let* $[w_0, w_1, w_2, w_3]$ *be the coordinates of any event relative to a coordinate frame having* $\underset{\sim}{Q}$ *as an origin in space and* $[Q_0]$ *as an origin in space-time.*

There is a non-singular 4×4 *matrix* $[a_{ij}]$ *such that*

(1) $$z_i = a_{ij}\, w_j \quad \text{and} \quad w_i = a^*_{ij}\, z_j \ ,$$

where $[z_0, z_1, z_2, z_3]$ *are the coordinates of the same event relative to a coordinate frame having* $\underset{\sim}{S}$ *as an origin in space and* $[S_0]$ *as an origin in space-time, and the coefficients of* $[a_{ij}]$ *satisfy the conditions:*

(2) $$\sum_{\alpha=1}^{3} a_{\alpha 0}^2 - a_{00}^2 = -1 \quad ,$$

$$\sum_{\alpha=1}^{3} a_{\alpha\beta}^2 - a_{0\beta}^2 = 1 \quad (\beta=1,2,3) \quad ,$$

$$\sum_{\alpha=1}^{3} a_{\alpha i}\, a_{\alpha k} - a_{0i}\, a_{0k} = 0 \quad (i,k=0,1,2,3 \text{ and } i \neq k) \quad ,$$

$$\sum_{\alpha=1}^{3} a_{0\alpha}^2 - a_{00}^2 = -1 \quad ,$$

$$\sum_{\alpha=1}^{3} a_{\beta\alpha}^2 - a_{\beta 0}^2 = 1 \quad (\beta=1,2,3) \quad ,$$

$$\sum_{\alpha=1}^{3} a_{i\alpha}\, a_{k\alpha} - a_{i0}\, a_{k0} = 0 \quad (i,k=0,1,2,3 \text{ and } i \neq k) \quad ,$$

$$det[a_{ij}] = \pm 1 \quad \text{and}$$

(3) $$a_{00} > 1 \quad .$$

This theorem is a consequence of Theorem 58 (§9.3), Corollaries 1 and 2 of Theorem 58 (§9.3), Theorem 61 (§9.5) and Corollary 3 of Theorem 61 (§9.5). It is used in the proof of Theorem 63 (§9.6).

PROOF. If we define a mapping τ_S as in Theorem 58 (§9.3), we see from Corollary 1 of the same theorem, that we can consistently define an analogous mapping τ_Q, since $\underset{\sim}{Q}$ and $\underset{\sim}{S}$ are synchronous. Now by Theorem 61 (§9.5), there is a coordinate frame which has $\underset{\sim}{S}$ as an origin in space and there is a coordinate frame which has $\underset{\sim}{Q}$ as an origin in space, both frames having the event $[S_0] = [Q_0]$ as the origin in space-time.

For events within the light cone whose vertex is $[Q_0]$ (= $[S_0]$), equations (1) and (2) are equivalent to equations (7), (2), (3) and (4) of §9.2. In order to establish equation (3), we observe that the first of equations (2) requires that $|a_{00}| \geqslant 1$; also a_{00} must be positive, since otherwise events within the upper half light cone would transform onto events within the lower half light cone. We have now established that the transformations (1) apply to the coordinates of events within the light cone whose vertex is $[Q_0]$.

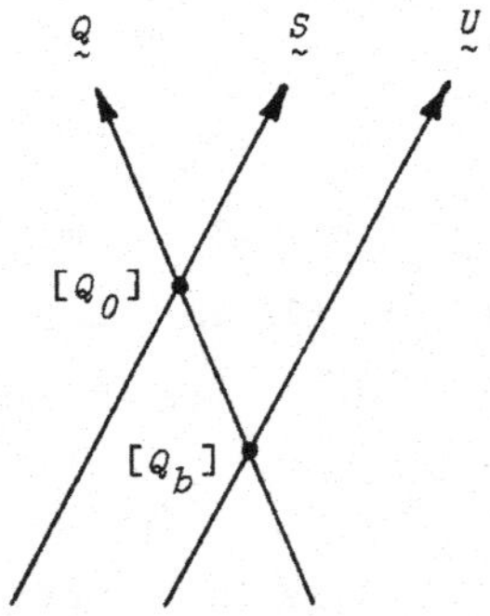

Fig. 58

In order to show that the transformations (1) apply to events which are not within the light cone whose vertex is $[Q_0]$, we take an arbitrary instant $Q_b \varepsilon \underset{\sim}{Q}$ with $Q_b < Q_0$ and consider the transformations which apply to events within the upper half light cone whose vertex is $[Q_b]$ (see Fig. 58). By Theorem 61 (§9.5) and Corollary 2 of Theorem 58 (§9.3), there is a particle $\underset{\sim}{U}$ such that:

(i) $\underset{\sim}{U}$ coincides with $\underset{\sim}{Q}$ at the event $[Q_b]$ (= $[U_0]$) ,

(ii) $\underset{\sim}{U} \parallel \underset{\sim}{S}$, and

(iii) $\underset{\sim}{U}$ is congruent to $\underset{\sim}{S}$ (see Fig. 58).

Then coordinate frames, having $\underset{\sim}{S}$ and $\underset{\sim}{U}$ as origins in space, have parallel position spaces so, by Corollary 3 of Theorem 61 (§9.5), there is a linear transformation of the form

$$\delta_1 : z_i \rightarrow x_i = c_i + z_i \quad ,$$

between coordinate frames having $\underset{\sim}{S}$ and $\underset{\sim}{U}$ as origins in space and $[Q_0]$ and $[Q_b]$, respectively, as origins in space-time. Again, by the same corollary, there is a linear transformation of the form

$$\delta_2 : y_i \rightarrow w_i = d_i + y_i \quad ,$$

between coordinate frames having origins in space which are permanently coincident with $\underset{\sim}{Q}$ and having $[Q_b]$ and $[Q_0]$ as origins in space-time. Now, as in the previous paragraph, there is a linear transformation of the form

$$\rho : x_i \to y_i = A^*_{ij}\, x_j \quad ,$$

between coordinate frames having $\underset{\sim}{U}$ and $\underset{\sim}{Q}$ as origins in space and $[Q_b]$ as an origin in space-time. Combining these three coordinate transformations, we obtain the transformation

$$\delta_2 \circ \rho \circ \delta_1 : z_i \to w_i = d_i + A^*_{ij}\, c_j + A^*_{ij}\, z_j$$

between the w-coordinate frame and the z-coordinate frame, which applies to all events within the light cone whose vertex is $[Q_b]$, and which must be identical with the transformation (1) within the upper half light cone whose vertex is $[Q_0]$. Since $[Q_b]$ is arbitrary and since each event is contained within some light cone having a vertex $[Q_c]$, for some instant $Q_c \in \underset{\sim}{Q}$, we see that the transformations (1) apply to all events. □

§9.7 Minkowski Space-Time

In this concluding section, we characterise the set of all coordinate frames by describing the coordinate transformations between them. The trajectories of particles and optical lines are then described relative to any coordinate frame.

THEOREM 63. *There is a maximal class of coordinate frames, $\mathcal{F}$, such that:*

(i) *$\mathcal{F}$ is the set of all coordinate frames which are related by coordinate transformations of the form*

(1) $$z_i = a_{ij}\, x_j - d_i \quad \text{and} \quad x_i = \overset{*}{a}_{ij}\, z_j + \overset{*}{a}_{ij}\, d_j \ ,$$

where $[d_0, d_1, d_2, d_3]$ is an arbitrary quadruple of real numbers and $[a_{ij}]$ is an arbitrary 4×4-matrix which satisfies the conditions:

(2) $$\sum_{\alpha=1}^{3} a_{\alpha 0}^2 - a_{00}^2 = -1 \ ,$$

$$\sum_{\alpha=1}^{3} a_{\alpha\beta}^2 - a_{0\beta}^2 = 1 \quad (\beta=1,2,3) \ ,$$

$$\sum_{\alpha=1}^{3} a_{\alpha i}\, a_{\alpha k} - a_{0i}\, a_{0k} = 0 \quad (i,k=0,1,2,3 \text{ and } i \neq k) \ ,$$

$$\sum_{\alpha=1}^{3} a_{0\alpha}^2 - a_{00}^2 = -1 \ ,$$

$$\sum_{\alpha=1}^{3} a_{\beta\alpha}^2 - a_{\beta 0}^2 = 1 \quad (\beta=1,2,3) \ ,$$

$$\sum_{\alpha=1}^{3} a_{i\alpha}\, a_{k\alpha} - a_{io}\, a_{ko} = 0 \quad (i,k=0,1,2,3 \text{ and } i \neq k) \ ,$$

$$\det[a_{ij}] = \pm 1 \ , \quad \text{and}$$

$$a_{00} \geq 1 \ ; \quad \text{and}$$

(ii) *given any frame not in* $\mathcal{F}$, *there is some frame in* $\mathcal{F}$ *and some positive real number* $\mu \neq 1$, *such that the two frames are related by a coordinate transformation of the form*

$$x_i = \mu\, w_i \quad . \tag{3}$$

REMARK. There are many classes of frames having the same properties as $\mathcal{F}$. This is easily demonstrated by choosing any positive real number $\mu \neq 1$ and applying the transformation (3) to all frames in $\mathcal{F}$.

This theorem is a consequence of Theorems 57 (§9.1), 59 (§9.4), 61 (§9.5) and Corollary 3 of 61 (§9.5) and Theorem 62 (§9.6). It is used in the proof of Theorem 64 (§9.7).

PROOF. It should be noted that the set of transformations (1) with $d_j = 0$ form a group (as explained in §9.2 above) which has a subgroup represented by matrices $[a_{ij}]$ for which

$$a_{00} = 1 \,, \quad a_{\beta 0} = 0 \,, \quad a_{0\beta} = 0$$

and therefore the submatrix $[a_{\alpha\beta}]$ is an orthogonal 3×3 matrix. We shall first prove a special case of (i) with

$$[d_0, d_1, d_2, d_3] = [0,0,0,0] \quad .$$

Take any particle $\underset{\sim}{Q}$ with a natural time scale; then for each particle $\underset{\sim}{T} \in SPR[Q_0]$ there is some particle $\underset{\sim}{S} \in \hat{\underset{\sim}{T}}$ such that $\underset{\sim}{S}$ and $\underset{\sim}{Q}$ are synchronous. By Theorem 59 (§9.4), this sub-SPRAY consists of an equivalence class of synchronous particles. By Corollary 3 of Theorem 61 (§9.5) we see that, to any given particle in this class, there is a set of coordinate frames, each having the given particle as an origin in space and the event $[Q_0]$ as an origin in space-time, whose coordinates are related by transformations of the form:

(4) $$y_0 = x_0 \ , \quad y_\alpha = b_{\alpha\beta}\, x_\beta \quad (\alpha=1,2,3) \ ,$$

where $[b_{\alpha\beta}]$ is any orthogonal 3×3 matrix. *We shall use the symbol* $\mathcal{S}$ *to denote the set of all coordinate frames defined in this way.* Then, by the previous theorem, for any two coordinate frames in $\mathcal{S}$, there is some 4×4 matrix $[a_{ij}]$ satisfying equations (2) such that the two coordinate frames are related by the transformations

(5) $$y_i = a_{ij}\, x_j \ .$$

Conversely, by Theorem 57 (§9.1), given any coordinate frame in $\mathcal{S}$ and any 4×4 matrix $[a_{ij}]$ satisfying equations (2), there is a particle $\underset{\sim}{T}$ which coincides with the set of events

(6) $$\{[\lambda a_{00}, -\lambda a_{01}, -\lambda a_{02}, -\lambda a_{03}] \ : \ -\infty < \lambda < \infty\}$$

which are specified in the given coordinate frame, which we shall call the x-coordinate frame. Take a particle $\underset{\sim}{S} \in \underset{\sim}{\hat{T}}$ such that $\underset{\sim}{Q}$ and $\underset{\sim}{S}$ are synchronous. Then Theorem 61 (§9.5) implies that there is a w-coordinate frame which has $\underset{\sim}{S}$ as an origin in space and $[S_0]$ (= $[Q_0]$) as an origin in space-time, and Theorem 62 (§9.6) implies that the two coordinate frames are related by transformations of the form

(7) $$w_k = b_{kj}\, x_j \quad and \quad x_k = b^*_{kj}\, w_j$$

where b^*_{ij} is the inverse of b_{ij} and

(8) $$b_{0j} = a_{0j}$$

in accordance with (6). We now define a matrix $[c_{ik}]$ such that

(9) $$c_{ik} \overset{def}{=} a_{ij}\, b^*_{jk}$$

and note that equations (6) of §9.2 and the fourth of equations (2) above imply that

(10) $$c_{00} = a_{0j}\, b^*_{j0} = a^2_{00} - \sum_{\alpha=1}^{3} a^2_{0\alpha} = 1 \quad .$$

Now by definition, both $[a_{ij}]$ and $[b^*_{ij}]$ satisfy equations having the form of equations (2) above. Matrices satisfying these equations form a group, as explained in §9.2, so the matrix $[c_{ij}]$ satisfies equations (2). Therefore equation (10) above and equations (15) of §9.2 imply that the submatrix $[c_{\alpha\beta}]$ is an orthogonal matrix. Corollary 3 of Theorem 61 (§9.5) implies that there is a y-coordinate frame which is related to the w-coordinate frame by the transformations

(11) $$y_i = c_{ik}\, w_k \quad .$$

Combining equations (7) and (11), we obtain equations (5), which completes the proof of the converse proposition. We have now proved a special case of (i) with

$$[d_0, d_1, d_2, d_3] = [0,0,0,0] \quad .$$

Now given any event and any coordinate frame in $\mathcal{Y}$, the given event has coordinates, say $[d_0, d_1, d_2, d_3]$ with respect to the coordinate frame, which we shall call the y-coordinate frame. By Corollary 3 to Theorem 61 (§9.5), there is a z-coordinate frame, whose origin in space-time is the given event and which is related to the y-coordinate frame by the transformations

(12) $$z_i = y_i - d_i \quad .$$

We shall use the symbol $\mathcal{Z}$ *to denote the set of all coordinate frames defined in this way.* Thus, combining equations having the form of (5) and (12), we find that any frame in $\mathcal{Z}$ is related to any frame in $\mathcal{Y}$ by equations having the form of (1). This set of equations forms a group and therefore any two frames in $\mathcal{Z}$ are related by a coordinate transformation having the form of (1).

Conversely, given an x-coordinate frame in $\mathcal{Z}$, an arbitrary quadruple of real numbers $[d_0, d_1, d_2, d_3]$ and an arbitrary 4×4 matrix $[a_{ij}]$ satisfying equations (2), there is a y-coordinate frame in $\mathcal{G}$ related to the x-coordinate frame in $\mathcal{F}$ by

(13) $$y_i = a_{ij}\, x_j - c_i$$

and, by definition of $\mathcal{F}$, there is a z-coordinate frame in $\mathcal{F}$ which is related to the y-coordinate frame in $\mathcal{G}$ by

(14) $$z_i = y_i + (c_i - d_i) \quad .$$

Combining equations (13) and (14), we obtain equations (1) which completes the proof of (i).

(ii) Given any coordinate frame, say the x-coordinate frame, there is some y-coordinate frame in $\mathcal{F}$ with the same origin in space-time. If these coordinate frames have synchronous origins in space, Theorem 62 (§9.6) implies that the x-coordinate frame is in $\mathcal{F}$. If the two frames do not have synchronous origins in space, Corollary 4 to Theorem 61 (§9.5) implies that there is some constant $\mu \neq 1$ and a w-coordinate frame, related to the given x-coordinate frame by the transformations (3), such that the origin in space of the w-coordinate frame is synchronous with the origin in space of the y-coordinate frame. Therefore the w-coordinate frame is in $\mathcal{F}$. This completes the proof of (ii) and shows that $\mathcal{F}$ is maximal. □

THEOREM 64 (Particle Trajectories)

Given a coordinate frame and a particle $\underset{\sim}{Q}$ *with a natural time scale, the coordinates of the events coincident with* $\underset{\sim}{Q}$ *are given by:*

(1) $$z_i(\lambda) = z_i(0) + \lambda v_i \quad (i=0,1,2,3) \quad , \quad \lambda \text{ real} ,$$

where

(2) $$v_0^2 - v_1^2 - v_2^2 - v_3^2 > 0 \quad \text{and} \quad v_0 > 0 \quad .$$

Conversely, given any coordinate frame and a set of events specified by the above conditions, there is a particle $\underset{\sim}{Q}$ *with a natural time scale such that the event whose coordinates are* $[z_i(\lambda) : i=0,1,2,3]$ *is* $[Q_\lambda]$ *where* $Q_\lambda \in \underset{\sim}{Q}$.

This theorem is a consequence of Theorems 61 (§9.5) and 63 (§9.7).

PROOF. The first proposition is a consequence of the previous theorem and Theorem 61 (§9.5): we can find a w-coordinate frame which has $\underset{\sim}{Q}$ as an origin in space and $[Q_0]$ as an origin in space-time, so for any instant $Q_\lambda \in \underset{\sim}{Q}$, the coordinates of the event $[Q_\lambda]$ are

$$w_i(\lambda) = \delta_{i0}\lambda \quad .$$

By combining the coordinate transformations (1) and (3) of the previous theorem, we obtain the equations

$$z_i(\lambda) = \mu a_{i0}\lambda - d_i \quad ,$$

which are equivalent to equations (1), with

$$v_i = \mu a_{i0} \quad and \quad z_i(0) = -d_i \quad .$$

The inequalities (2) are consequences of (the first equation and the last inequality of (2) of) the previous theorem.

Conversely, given equations (1) and (2), we can find a matrix $[a_{ij}]$ and define $[d_i : i=0,1,2,3]$ and μ, as in the previous theorem, such that

(3) $$a_{i0} = \mu^{-1} v_i \quad and \quad d_i = -z_i(0)$$

where $\mu = (v_0^2 - v_1^2 - v_2^2 - v_3^2)^{\frac{1}{2}}$.

By the previous theorem, there is an x-coordinate frame which is related to the z-coordinate frame by equations (1) of the previous theorem, and there is a w-coordinate frame which is related to the x-coordinate frame by equations (3) of the previous theorem. Combining these coördinate transformations, we obtain.

$$z_i = \mu a_{ij} \, w_j - d_i \quad .$$

Inverting these equations and substituting from (1) and (3), we find that

$$w_k(\lambda) = \delta_{k0}\lambda \quad .$$

Now let $\underset{\sim}{Q}$ be an origin in space of the w-coordinate frame. □

THEOREM 65 (Optical Lines)

Given a coordinate frame and an optical line, there are quadruples of real numbers $[x'_0, x'_1, x'_2, x'_3]$ *and* $[c_0, c_1, c_2, c_3]$ *such that the coordinates of all events, corresponding to instants of the optical line, are given by the equations*

(1) $$(x_i(\lambda) - x'_i)/c_i = \lambda \quad ,$$

where λ *is a real variable and*

(2) $$c_0^2 - c_1^2 - c_2^2 - c_3^2 = 0 \quad .$$

Conversely, given quadruples $[x'_0, x'_1, x'_2, x'_3]$ *and* $[c_0, c_1, c_2, c_3]$ *satisfying (2), there is an optical line whose instants are elements of the events specified by equations (1).*

PROOF. Both propositions are consequences of equations (2) of Theorem 61 (§9.5). □

CHAPTER 10

CONCLUDING REMARKS

Our task is now complete in that we have described Minkowski space-time in terms of undefined elements called "particles" and a single undefined "signal relation". We have demonstrated that our axiom system is categoric for Minkowski space-time. However, as mentioned in the introduction, we have not discussed the question of independence of the axioms. It is quite likely that there is some interdependence between the axioms and that the axiom system could be improved by the substitution of weaker axioms. However the author is aware of counterexamples which can be used to demonstrate the independence of some of the axioms; namely, Axioms VII (§2.9), VIII (§2.10), X (§2.12) and XI (§2.13): also certain subsets of the other axioms can be shown to be independent from those remaining. Consequently, the possibilities for modification of the axioms are subject to a number of constraints.

Minkowski space-time is a pseudo-euclidean space of signature (+, +, +, -). In many ways, the de Sitter universe is the corresponding analogue of the non-euclidean hyperbolic space and, in the present context, the most relevant analogy is that parallelism is not unique in these spaces. The present system of axioms can be modified so as to be valid propositions in a de Sitter universe: the principal alteration is to the Signal Axiom (Axiom I, §2.2) which must be modified to take the de Sitter "event horizon" into account. All but one of the remaining axioms can be altered slightly so as to be in accordance with the new Signal Axiom; the only exception being the Axiom of Connectedness (Axiom X, §2.12) which can be re-expressed in two different forms to correspond to the two non-isomorphic models of the de Sitter universe which are discussed by Schrödinger [1956]. Whether or not these propositions form categorical axiom systems is a question which remains to be investigated.

Finally, we remark that other directions for investigation of more general space-time structures have been described by Busemann [1967] and Pimenov [1970], who have applied topological axioms to ordered structures called "space-times" whose undefined elements are called "events". Both of these authors aim at extending our knowledge of possible space-time structures. They express space-time theory in terms of a single relation (before-after) and so their approaches are more akin to that of Robb [1921, 1936] rather than to that of Walker [1948, 1959]. Their methods have much in common with those of geometry and topology.

APPENDIX 1

CHARACTERISATION OF THE ELEMENTARY SPACES

In the present treatment we are interested in showing that each SPRAY is a hyperbolic space of three dimensions, for this property of each SPRAY is intimately related to the Lorentz transformation formulae (see §9.6). The problem of characterising hyperbolic spaces is a special case of the famous "Riemann-Helmholtz" or "Helmholtz-Lie" problem which is reviewed by Freudenthal [1965]. A recent characterisation by Tits [1953, 1955] is used in the present treatment for two reasons: firstly the dimension of repidity space need not be assumed, and secondly the "double transitivity" of the motions of rapidity space is a consequence of the Axiom of Isotropy of SPRAYs (Axiom VII, §2.9). The characterisation by Tits and its proof are discussed by Busemann [1955, 1970].

Given a non-empty set X, a collection of subsets $\mathfrak{J}$ is a *topology* on X if:

(i) ϕ, $X \varepsilon \mathfrak{J}$ (ϕ is the empty set),

(ii) the union of every class of sets in $\mathfrak{J}$ is a set in $\mathfrak{J}$, and

(iii) the intersection of every finite class of sets in $\mathfrak{J}$ is a set in $\mathfrak{J}$.

The sets in the class $\mathcal{J}$ are called the *open sets* of the topological space $(X, \mathcal{J})$ and the elements of X are called its *points*. A *closed set* in a topological space is a set whose complement is open. A *neighbourhood* of a point in a topological space is an open set which contains the point. If A is a subset of a topological space, the *closure* of A is the intersection of all closed sets which contain A.

A topological space is said to be *connected* if the only two subsets which are both open and closed are X and ϕ.

A class $\{O_i\}$ of open subsets of X is said to be an *open cover* of X if each point in X belongs to at least one O_i; that is, if $\bigcup_i O_i = X$. A subclass of an open cover which is itself an open cover is called a *subcover*. A *compact space* is a topological space in which every open cover has a finite subcover. A topological space is *locally compact* if each of its points has a neighbourhood with compact closure.

A set X is called a *metric space* if to each pair of elements $x, y \in X$ there is a real number $d(x, y) \geq 0$, called the *distance* between x and y, which satisfies the following three conditions:

(i) $d(x, y) = 0$ *if and only if* $x = y$,

(ii) $d(x, y) = d(y, x)$, and

(iii) *for any* $x, y, z \in X$, $d(x, y) + d(y, z) \geq d(x, z)$.

The *diameter* of X is $\sup_{x,y\in X} d(x, y)$. A metric space X is *bounded* if it has a finite diameter.

A *curve* in X is a continuous mapping of a closed interval of the reals into X. The *length* $\lambda(x)$ *of a curve* $x(t)$ $(\alpha \leqslant t \leqslant \beta)$ is defined in the usual way: for any partition $\Delta: \alpha = t_0 < t_1 < \cdots < t_k = \beta$ we put

$$\lambda(x, \Delta) = \sum_{i=1}^{k} d\left[x(t_{i-1}), x(t_i)\right] \text{ and } \lambda(x) \overset{\text{def}}{=} \sup_{\Delta} \lambda(x, \Delta).$$

Then for each Δ,

$$d(x(\alpha), x(\beta)) \leqslant \lambda(x, \Delta) \leqslant \lambda(x).$$

The space X is *arcwise connected* if, for any two points x and y, there is a curve whose end-points are x and y.

A metric function d on X is *intrinsic* if for any pair of points x and y, the distance $d(x, y)$ is equal to the infimum of the lengths of all curves from x to y. This concept presupposes that X is arcwise connected.

A *motion* of the space X is a mapping of X onto itself which preserves distances: that is, a motion is an isometric mapping.

A space X has a *doubly transitive group of motions* if, for any two ordered point pairs with equal distances, there is a motion which maps the first pair into the second pair.

A space X is *isotropic at a point* $x \in X$ if, for any two points $y, z \in X$ such that $d(x, y) = d(x, z)$, there is a motion which sends y into z and leaves x invariant. If the space X is isotropic at all its points, we will simply say that X is *isotropic*, or that X is an *isotropic space*.

LEMMA

An arcwise connected topological space is connected.

PROOF

This is a well-known result: see, for example, Mendelson [1962, §4.6].

LEMMA

An arcwise connected metric space is isotropic if and only if it has a doubly transitive group of motions.

PROOF

Let x, y, z be three points in X such that

$$d(x, y) = d(x, z).$$

Then the ordered point pairs (x, y) and (x, z) have equal distances and so double transitivity implies that there is a motion which sends the ordered pair (x, y) onto the ordered pair (x, z).

Appendix 1]

Let x, x', y, y' be any four points in X with

$$d(x, y) = d(x', y').$$

Since X is arcwise connected, there is a curve from x to x'. If we take an arbitrary point w on this curve, the functions $d(w, x)$ and $d(w, x')$ are continuous real-valued functions of arc length along the curve, so by the Intermediate Value Theorem of real variable theory (see Fulks [1961]), there is a point z on the curve such that

$$d(z, x) = d(z, x').$$

Now since the space X is isotropic, there is a motion which sends x onto x' and y onto some point y'' such that

$$d(x,y) = d(x',y'') = d(x',y')$$

Now there is an isotropy mapping about x' which sends y'' onto y''. Since the composition of two motions is a motion, we have shown that there is a motion which sends the ordered pair (x, y) onto the ordered pair (x', y'). □

THEOREM (Tits [1952, 1955])

If X is a locally compact connected metric space with a doubly transitive group of motions, then X is finite-dimensional and is either an elliptic, euclidean or hyperbolic space, or an elliptic or hyperbolic hermitian or quaternion space, or an elliptic or hyperbolic Cayley plane; with the reservation that the distance may be of the form $g(d(x, y))$ where $d(x, y)$ is an intrinsic distance function on the space.

These spaces are described by Busemann [1955, §53] and Tits [1952; 1955 §II.E.]. Of the above spaces the only ones which are unbounded are the euclidean spaces and the various types of hyperbolic spaces. Furthermore, it is known that a maximal set of equidistant points has exactly 4 members in the euclidean and hyperbolic spaces of 3 dimensions. The hyperbolic hermitian spaces have dimension $2N$ and the hyperbolic quaternion spaces have dimension $4N$, where N is any positive integer. A set of equidistant points has at most 3 members in the hyperbolic hermitian space of dimension 2, at least 5 members in the hyperbolic hermitian spaces of dimension $2N (N \geq 2)$ and in all of the hyperbolic quaternion spaces and hence, by comparison with the hyperbolic quaternion space of 4 dimensions, in the hyperbolic Cayley plane.

We conclude this section by summarising the previous remarks, the theorem of Tits, and the preceding two lemmas in the form:

If X is an unbounded locally compact arcwise-connected isotropic metric space such that any maximal set of equidistant points has 4 members, then X is either a euclidean or a hyperbolic space of 3 dimensions.

APPENDIX 2

HOMOGENEOUS COORDINATES IN HYPERBOLIC AND EUCLIDEAN SPACES

(i) Projective n-Space (see Busemann and Kelly [1953])

We first discuss homogeneous coordinates in a projective space of n dimensions, where n is a positive integer. Let $x = (x_0, x_1, \ldots, x_n)$ and $y = (y_0, y_1, \ldots, y_n)$ be $(n+1)$-tuples of real numbers (not all zero). If the $(n+1)$-tuples are proportional, x and y are said to be members of the same *class*. *Points* in projective n-space can be represented by *classes of $(n+1)$-tuples*. A set of points $\{x^1, x^2, \ldots, x^m\}$ is *independent* if the set of $(n+1)$-tuples representing them is linearly independent. The maximum number of linearly independent classes of $(n+1)$-tuples is $(n+1)$.

If $\bar{p}^0, \bar{p}^1, \ldots, \bar{p}^n$ are given representations of $n + 1$ independent points, and $\bar{x}$ is a given representation of an arbitrary point, then the equations

$$\bar{x}_k = \sum_{i=0}^{n} x_i^* \bar{p}_k^i \qquad (k=0,1,\ldots,n)$$

determine the $\{x_i^*\}$ uniquely, since the matrix $[\bar{p}_k^i]$ is non-singular, due to the $\{\bar{p}_i\}$ being independent. However to specify the point $\bar{x}$ it is only necessary to specify the

$\{x_i^*\}$ to within an arbitrary non-zero multiplicative factor: the $\{x_i^*\}$ are called *projective coordinates* of $\bar{x}$ relative to the basis $\{\bar{p}^i\}$. If we define a basis

$$p^i = (\delta_i^0, \delta_i^1, \ldots, \delta_i^n),$$

where δ_i^j is the Kronecker delta, the corresponding projective coordinates are called *special projective coordinates*. A *change of basis* results in a linear non-singular transformation of projective coordinates.

Three points x, y, z are collinear if there are real numbers a and b such that for all $i \in \{0, 1, \ldots, n\}$

$$z_i = ax_i + by_i.$$

(ii) n-Dimensional Hyperbolic Geometry

All (real) hyperbolic geometries of the same dimension and curvature are isometric. A model of n-dimensional hyperbolic geometry which has a direct relevance to the relationships between rapidity, velocity, coordinate distance and coordinate time is the Hilbert geometry whose domain is the interior of the unit n-sphere

$$E: (x_1')^2 + (x_2')^2 + \ldots + (x_n')^2 \leqslant 1,$$

where $x_1' = x_1/x_0, \ldots, x_n' = x_n/x_0$ are special projective coordinates. From the inequality above

$$x_1^2 + x_2^2 + \ldots + x_n^2 - x_0^2 \leqslant 0.$$

Appendix 2(iii)]

With $c = (1, 0, \ldots, 0)$ and $x = (x_0, x_1, \ldots, x_n)$, the hyperbolic distance $h(x, c)$ is related to the euclidean distance between x and the centre c by

(1) $$[(x_1')^2 + \ldots + (x_n')^2]^{\frac{1}{2}} = \tanh h(x, c)/k,$$

where k is a positive constant and the curvature of the space is $-k$. More generally, if we define, for any two points x and y,

$$\Omega(x, y) \overset{def}{=} -x_0 y_0 + x_1 y_1 + x_2 y_2 + \ldots + x_n y_n,$$

then

(2) $$h(x,y) = k \, Arcosh\{|\Omega(x,y)|[\Omega(x,x)\Omega(y,y)]^{-\frac{1}{2}}\}$$

(iii) n-Dimensional Euclidean Geometry

A model of n-dimensional euclidean geometry, which has relevance to the relationships between velocity, distance and time in Newtonian kinematics (but not in the kinematics of special relativity) is the geometry whose domain is

$$E: (x_1')^2 + (x_2')^2 + \ldots + (x_n')^2 < \infty,$$

where $x_1' = x_1/x_0, \ldots, x_n' = x_n/x$ are special projective coordinates. With $c = (1, 0, \ldots, 0)$ and $x = (x_0, \ldots, x_n)$, the euclidean distance $e(x, c)$ is given by

(3) $$e(x, c) = [(x_1')^2 + \ldots + (x_n')^2]^{\frac{1}{2}}.$$

BIBLIOGRAPHY

Figures in square brackets, which follow each reference, indicate the sections where the reference is cited.

J. Aczèl, [1966] Functional Equations and their Applications, (Academic Press). [§9.1].

G. Birkhoff [1967] Lattice Theory, 3rd edition. American Math. Soc. Colloquium Publications, vol. 25. Providence, Rhode Island. [§3.1].

J. Bolyai [1832] Appendix scientiam spatii absolute veram exhibens: a veritate aut falsitate Axiomatis XI Euclidei (a priori haud unquam decidenda) independentem. See translation in Bonola [1955]. [§1.0].

R. Bonola [1955] Non-euclidean Geometry (Dover). [§1.0].

R.H. Boyer [1965] Some uses of hyperbolic velocity space. Amer. J. Phys., 33, 910 - 916.

M. Bunge [1967] Foundations of Physics (Springer-Verlag). [§1.0].

H. Busemann and P.J. Kelly [1953] Projective Geometry and Projective Metrics (Academic Press). [Appendix 2].

H. Busemann [1955] The Geometry of Geodesics (Academic Press, N.Y.). [§1.0, 2.7, 2.13, Appendix 1].

H. Busemann [1967] Timelike spaces, Rozpr. Mat. 53. [§10.0].

H. Busemann [1970] Recent Synthetic Differential Geometry, Ergebnisse der Mathematik und ihrer Grenzgebiete Band 54 (Springer-Verlag). [Appendix 1].

A. Einstein [1905] Zur Elektrodynamik bewegter Körper, Ann. der Phys. vol. 17, 891. [§1.0, 2.9].

. Euclid [~ 300 B.C.]. See T.L. Heath [1956] The Thirteen Books of Euclid's Elements (Dover) also (Cambridge U.P., 1908). [§1.0, 2.12].

V. Fock [1964] The theory of space, time and gravitation. Translated by N. Kemmer, second revised edition (Pergamon, 1964).

H. Freudenthal [1965] Lie Groups in the Foundations of Geometry. Advances in Math. 1, 145-190. [Appendix 1].

W. Fulks [1961] Advanced Calculus (Wiley). [§5.2, 7.3, Appendix 1].

N.I. Lobachevsky [1829] Études Géométriques sur la Théorie des Parallels. See translation in Bonola [1955]. [§1.0].

H. MacNeille [1937] Partially ordered sets. Trans. Amer. Math. Soc. 42, 416-460. [§3.1].

B. Mendelson [1962] Introduction to Topology (Allyn and Bacon) [Appendix 1].

E.A. Milne [1948] Kinematic Relativity (Oxford). [§8.1].

W. Noll [1964] Euclidean geometry and Minkowskian chronometry. Amer. Math. Monthly 71, 129-144. [§1.0].

W. Pauli [1921] Theory of relativity. Translated from the German by G. Field (Pergamon, 1958).

R.I. Pimenov [1970] Kinematic Spaces, Seminars in Mathematics, vol. 6, Steklov Mathematical Institute, Leningrad. Translated from Russian by Consultants Bureau, New York. [§10.0].

A.V. Pogorolev [1966] Lectures on the Foundations of Geometry, translated from the second Russian edition by L.F. Boron and W.D. Bouwsma (Noordhoff). [§2.9].

A.A. Robb [1921] The Absolute Relations of Time and Space (Cambridge U.P.). [§8.1].

A.A. Robb [1936] Geometry of Time and Space (Cambridge U.P.). [§1.0].

H. Rubin and J.E. Rubin [1963] Equivalents of the Axiom of Choice (North Holland, Amsterdam). [§5.2].

E. Schrödinger [1956] Expanding Universes (Cambridge U.P.). [§10.0].

W. Sierpinski [1965] Cardinal and Ordinal Numbers. Second edition revised (Warszawa). [§7.3].

J.A. Smorodinsky [1965] Kinematik and Lobatschewski-geometrie. Fortschr. Physik, 13, 157-173.

P. Suppes and H. Rubin [1954] Transformations of systems of relativistic particle mechanics. Pacific Journal of Math. vol. 4, 563-601. [§1.0].

P. Suppes [1959] Axioms for Relativistic Kinematics With or Without Parity, The axiomatic method. With special reference to geometry and physics (Proc. International Symposium, University of California, Berkeley, December 1957 - January 1958, 291-307. North Holland, Amsterdam, 1959). [§1.0,2.2,3.0].

G. Szekeres [1968] Kinematic Geometry: An Axiomatic System for Minkowski Space-time. J. Aust. Math. Soc., vol. VIII, 134-160. [§1.0,2.2,2.9,7.3,8.1].

J. Tits [1952] Étude de certains espaces métriques. Bull. Soc. Math. Belgique, 44-52, c.f. also: Sur un article précédent, same bulletin, 1953, 126-127. [§1.0, Appendix 1].

J. Tits [1955] Sur certaines classes d'espaces homogènes de groups de Lie. Mem. Acad. Roy. Belg. Sci., 29, fasc. 3. [§1.0, Appendix 1].

A.G. Walker [1948] Foundations of Relativity: Parts I and II, Proc. Roy. Soc. Edinburgh Sect A. 62, 319-335. [§1.0,2.1,2.2,2.3,2.4,2.5,2.6,2.7,3.0,3.2,3.3,3.4, 3.6,4.1,4.2,4.3,7.3].

A.G. Walker [1959] "Axioms for cosmology", The axiomatic method. With special reference to geometry and physics (Proc. International Symposium, University of California, Berkeley, December 1957 - January 1958, 308-321. North Holland, Amsterdam, 1959). [§1.0,2.2,3.0].

E.C. Zeeman [1964] Causality implies the Lorentz group. J. Math. Phys., 5, 490-493. [§1.0].

Figures in square brackets, which follow each reference, indicate the sections where the reference is cited.

Vol. 215: P. Antonelli, D. Burghelea and P. J. Kahn, The Concordance-Homotopy Groups of Geometric Automorphism Groups. X, 140 pages. 1971. DM 16,-

Vol. 216: H. Maaß, Siegel's Modular Forms and Dirichlet Series. VII, 328 pages. 1971. DM 20,-

Vol. 217: T. J. Jech, Lectures in Set Theory with Particular Emphasis on the Method of Forcing. V, 137 pages. 1971. DM 16,-

Vol. 218: C. P. Schnorr, Zufälligkeit und Wahrscheinlichkeit. IV, 212 Seiten. 1971. DM 20,-

Vol. 219: N. L. Alling and N. Greenleaf, Foundations of the Theory of Klein Surfaces. IX, 117 pages. 1971. DM 16,-

Vol. 220: W. A. Coppel, Disconjugacy. V, 148 pages. 1971. DM 16,-

Vol. 221: P. Gabriel und F. Ulmer, Lokal präsentierbare Kategorien. V, 200 Seiten. 1971. DM 18,-

Vol. 222: C. Meghea, Compactification des Espaces Harmoniques. III, 108 pages. 1971. DM 16,-

Vol. 223: U. Felgner, Models of ZF-Set Theory. VI, 173 pages. 1971. DM 16,-

Vol. 224: Revètements Etales et Groupe Fondamental. (SGA 1). Dirigé par A. Grothendieck XXII, 447 pages. 1971. DM 30,-

Vol. 225: Théorie des Intersections et Théorème de Riemann-Roch. (SGA 6). Dirigé par P. Berthelot, A. Grothendieck et L. Illusie. XII, 700 pages. 1971. DM 40,-

Vol. 226: Seminar on Potential Theory, II. Edited by H. Bauer. IV, 170 pages. 1971. DM 18,-

Vol. 227: H. L. Montgomery, Topics in Multiplicative Number Theory. IX, 178 pages. 1971. DM 18,-

Vol. 228: Conference on Applications of Numerical Analysis. Edited by J. Ll. Morris. X, 358 pages. 1971. DM 26,-

Vol. 229: J. Väisälä, Lectures on n-Dimensional Quasiconformal Mappings. XIV, 144 pages. 1971. DM 16,-

Vol. 230: L. Waelbroeck, Topological Vector Spaces and Algebras. VII, 158 pages. 1971. DM 16,-

Vol. 231: H. Reiter, L^1-Algebras and Segal Algebras. XI, 113 pages. 1971. DM 16,-

Vol. 232: T. H. Ganelius, Tauberian Remainder Theorems. VI, 75 pages. 1971. DM 16,-

Vol. 233: C. P. Tsokos and W. J. Padgett. Random Integral Equations with Applications to stochastic Systems. VII, 174 pages. 1971. DM 18,-

Vol. 234: A. Andreotti and W. Stoll. Analytic and Algebraic Dependence of Meromorphic Functions. III, 390 pages. 1971. DM 26,-

Vol. 235: Global Differentiable Dynamics. Edited by O. Hájek, A. J. Lohwater, and R. McCann. X, 140 pages. 1971. DM 16,-

Vol. 236: M. Barr, P. A. Grillet, and D. H. van Osdol. Exact Categories and Categories of Sheaves. VII, 239 pages. 1971. DM 20,-

Vol. 237: B. Stenström, Rings and Modules of Quotients. VII, 136 pages. 1971. DM 16,-

Vol. 238: Der kanonische Modul eines Cohen-Macaulay-Rings. Herausgegeben von Jürgen Herzog und Ernst Kunz. VI, 103 Seiten. 1971. DM 16,-

Vol. 239: L. Illusie, Complexe Cotangent et Déformations I. XV, 355 pages. 1971. DM 26,-

Vol. 240: A. Kerber, Representations of Permutation Groups I. VII, 192 pages. 1971. DM 18,-

Vol. 241: S. Kaneyuki, Homogeneous Bounded Domains and Siegel Domains. V, 89 pages. 1971. DM 16,-

Vol. 242: R. R. Coifman et G. Weiss, Analyse Harmonique Non-Commutative sur Certains Espaces. V, 160 pages. 1971. DM 16,-

Vol. 243: Japan-United States Seminar on Ordinary Differential and Functional Equations. Edited by M. Urabe. VIII, 332 pages. 1971. DM 26,-

Vol. 244: Séminaire Bourbaki – vol. 1970/71. Exposés 382–399. IV, 356 pages. 1971. DM 26,-

Vol. 245: D. E. Cohen, Groups of Cohomological Dimension One. V, 99 pages. 1972. DM 16,-

Vol. 246: Lectures on Rings and Modules. Tulane University Ring and Operator Theory Year, 1970–1971. Volume I. X, 661 pages. 1972. DM 40,-

Vol. 247: Lectures on Operator Algebras. Tulane University Ring and Operator Theory Year, 1970–1971. Volume II. XI, 786 pages. 1972. DM 40,-

Vol. 248: Lectures on the Applications of Sheaves to Ring Theory. Tulane University Ring and Operator Theory Year, 1970–1971. Volume III. VIII, 315 pages. 1971. DM 26,-

Vol. 249: Symposium on Algebraic Topology. Edited by P. J. Hilton. VII, 111 pages. 1971. DM 16,-

Vol. 250: B. Jónsson, Topics in Universal Algebra. VI, 220 pages. 1972. DM 20,-

Vol. 251: The Theory of Arithmetic Functions. Edited by A. A. Gioia and D. L. Goldsmith VI, 287 pages. 1972. DM 24,-

Vol. 252: D. A. Stone, Stratified Polyhedra. IX, 193 pages. 1972. DM 18,-

Vol. 253: V. Komkov, Optimal Control Theory for the Damping of Vibrations of Simple Elastic Systems. V, 240 pages. 1972. DM 20,-

Vol. 254: C. U. Jensen, Les Foncteurs Dérivés de $\varprojlim$ et leurs Applications en Théorie des Modules. V, 103 pages. 1972. DM 16,-

Vol. 255: Conference in Mathematical Logic – London '70. Edited by W. Hodges. VIII, 351 pages. 1972. DM 26,-

Vol. 256: C. A. Berenstein and M. A. Dostal, Analytically Uniform Spaces and their Applications to Convolution Equations. VII, 130 pages. 1972. DM 16,-

Vol. 257: R. B. Holmes, A Course on Optimization and Best Approximation. VIII, 233 pages. 1972. DM 20,-

Vol. 258: Séminaire de Probabilités VI. Edited by P. A. Meyer. VI, 253 pages. 1972. DM 22,-

Vol. 259: N. Moulis, Structures de Fredholm sur les Variétés Hilbertiennes. V, 123 pages. 1972. DM 16,-

Vol. 260: R. Godement and H. Jacquet, Zeta Functions of Simple Algebras. IX, 188 pages. 1972. DM 18,-

Vol. 261: A. Guichardet, Symmetric Hilbert Spaces and Related Topics. V, 197 pages. 1972. DM 18,-

Vol. 262: H. G. Zimmer, Computational Problems, Methods, and Results in Algebraic Number Theory. V, 103 pages. 1972. DM 16,-

Vol. 263: T. Parthasarathy, Selection Theorems and their Applications. VII, 101 pages. 1972. DM 16,-

Vol. 264: W. Messing, The Crystals Associated to Barsotti-Tate Groups: With Applications to Abelian Schemes. III, 190 pages. 1972. DM 18,-

Vol. 265: N. Saavedra Rivano, Catégories Tannakiennes. II, 418 pages. 1972. DM 26,-

Vol. 266: Conference on Harmonic Analysis. Edited by D. Gulick and R. L. Lipsman. VI, 323 pages. 1972. DM 24,-

Vol. 267: Numerische Lösung nichtlinearer partieller Differential- und Integro-Differentialgleichungen. Herausgegeben von R. Ansorge und W. Törnig, VI, 339 Seiten. 1972. DM 26,-

Vol. 268: C. G. Simader, On Dirichlet's Boundary Value Problem. IV, 238 pages. 1972. DM 20,-

Vol. 269: Théorie des Topos et Cohomologie Etale des Schémas. (SGA 4). Dirigé par M. Artin, A. Grothendieck et J. L. Verdier. XIX, 525 pages. 1972. DM 50,-

Vol. 270: Théorie des Topos et Cohomologie Etale des Schémas. Tome 2. (SGA 4). Dirigé par M. Artin, A. Grothendieck et J. L. Verdier. V, 418 pages. 1972. DM 50,-

Vol. 271: J. P. May, The Geometry of Iterated Loop Spaces. IX, 175 pages. 1972. DM 18,-

Vol. 272: K. R. Parthasarathy and K. Schmidt, Positive Definite Kernels, Continuous Tensor Products, and Central Limit Theorems of Probability Theory. VI, 107 pages. 1972. DM 16,-

Vol. 273: U. Seip, Kompakt erzeugte Vektorräume und Analysis. IX, 119 Seiten. 1972. DM 16,-

Vol. 274: Toposes, Algebraic Geometry and Logic. Edited by. F. W. Lawvere. VI, 189 pages. 1972. DM 18,-

Vol. 275: Séminaire Pierre Lelong (Analyse) Année 1970–1971. VI, 181 pages. 1972. DM 18,-

Vol. 276: A. Borel, Représentations de Groupes Localement Compacts. V, 98 pages. 1972. DM 16,-

Vol. 277: Séminaire Banach. Edité par C. Houzel. VII, 229 pages. 1972. DM 20,-

Vol. 278: H. Jacquet, Automorphic Forms on GL(2). Part II. XIII, 142 pages. 1972. DM 16,–

Vol. 279: R. Bott, S. Gitler and I. M. James, Lectures on Algebraic and Differential Topology. V, 174 pages. 1972. DM 18,–

Vol. 280: Conference on the Theory of Ordinary and Partial Differential Equations. Edited by W. N. Everitt and B. D. Sleeman. XV, 367 pages. 1972. DM 26,–

Vol. 281: Coherence in Categories. Edited by S. Mac Lane. VII, 235 pages. 1972. DM 20,–

Vol. 282: W. Klingenberg und P. Flaschel, Riemannsche Hilbertmannigfaltigkeiten. Periodische Geodätische. VII, 211 Seiten. 1972. DM 20,–

Vol. 283: L. Illusie, Complexe Cotangent et Déformations II. VII, 304 pages. 1972. DM 24,–

Vol. 284: P. A. Meyer, Martingales and Stochastic Integrals I. VI, 89 pages. 1972. DM 16,–

Vol. 285: P. de la Harpe, Classical Banach-Lie Algebras and Banach-Lie Groups of Operators in Hilbert Space. III, 160 pages. 1972. DM 16,–

Vol. 286: S. Murakami, On Automorphisms of Siegel Domains. V, 95 pages. 1972. DM 16,–

Vol. 287: Hyperfunctions and Pseudo-Differential Equations. Edited by H. Komatsu. VII, 529 pages. 1973. DM 36,–

Vol. 288: Groupes de Monodromie en Géométrie Algébrique. (SGA 7 I). Dirigé par A. Grothendieck. IX, 523 pages. 1972. DM 50,–

Vol. 289: B. Fuglede, Finely Harmonic Functions. III, 188. 1972. DM 18,–

Vol. 290: D. B. Zagier, Equivariant Pontrjagin Classes and Applications to Orbit Spaces. IX, 130 pages. 1972. DM 16,–

Vol. 291: P. Orlik, Seifert Manifolds. VIII, 155 pages. 1972. DM 16,–

Vol. 292: W. D. Wallis, A. P. Street and J. S. Wallis, Combinatorics: Room Squares, Sum-Free Sets, Hadamard Matrices. V, 508 pages. 1972. DM 50,–

Vol. 293: R. A. DeVore, The Approximation of Continuous Functions by Positive Linear Operators. VIII, 289 pages. 1972. DM 24,–

Vol. 294: Stability of Stochastic Dynamical Systems. Edited by R. F. Curtain. IX, 332 pages. 1972. DM 26,–

Vol. 295: C. Dellacherie, Ensembles Analytiques, Capacités, Mesures de Hausdorff. XII, 123 pages. 1972. DM 16,–

Vol. 296: Probability and Information Theory II. Edited by M. Behara, K. Krickeberg and J. Wolfowitz. V, 223 pages. 1973. DM 20,–

Vol. 297: J. Garnett, Analytic Capacity and Measure. IV, 138 pages. 1972. DM 16,–

Vol. 298: Proceedings of the Second Conference on Compact Transformation Groups. Part 1. XIII, 453 pages. 1972. DM 32,–

Vol. 299: Proceedings of the Second Conference on Compact Transformation Groups. Part 2. XIV, 327 pages. 1972. DM 26,–

Vol. 300: P. Eymard, Moyennes Invariantes et Représentations Unitaires. II. 113 pages. 1972. DM 16,–

Vol. 301: F. Pittnauer, Vorlesungen über asymptotische Reihen. VI, 186 Seiten. 1972. DM 18,–

Vol. 302: M. Demazure, Lectures on p-Divisible Groups. V, 98 pages. 1972. DM 16,–

Vol. 303: Graph Theory and Applications. Edited by Y. Alavi, D. R. Lick and A. T. White. IX, 329 pages. 1972. DM 26,–

Vol. 304: A. K. Bousfield and D. M. Kan, Homotopy Limits, Completions and Localizations. V, 348 pages. 1972. DM 26,–

Vol. 305: Théorie des Topos et Cohomologie Etale des Schémas. Tome 3. (SGA 4). Dirigé par M. Artin, A. Grothendieck et J. L. Verdier. VI, 640 pages. 1973. DM 50,–

Vol. 306: H. Luckhardt, Extensional Gödel Functional Interpretation. VI, 161 pages. 1973. DM 18,–

Vol. 307: J. L. Bretagnolle, S. D. Chatterji et P.-A. Meyer, Ecole d'été de Probabilités: Processus Stochastiques. VI, 198 pages. 1973. DM 20,–

Vol. 308: D. Knutson, λ-Rings and the Representation Theory of the Symmetric Group. IV, 203 pages. 1973. DM 20,–

Vol. 309: D. H. Sattinger, Topics in Stability and Bifurcation Theory. VI, 190 pages. 1973. DM 18,–

Vol. 310: B. Iversen, Generic Local Structure of the Morphisms in Commutative Algebra. IV, 108 pages. 1973. DM 16,–

Vol. 311: Conference on Commutative Algebra. Edited by J. W. Brewer and E. A. Rutter. VII, 251 pages. 1973. DM 22,–

Vol. 312: Symposium on Ordinary Differential Equations. Edited by W. A. Harris, Jr. and Y. Sibuya. VIII, 204 pages. 1973. DM 22,–

Vol. 313: K. Jörgens and J. Weidmann, Spectral Properties of Hamiltonian Operators. III, 140 pages. 1973. DM 16,–

Vol. 314: M. Deuring, Lectures on the Theory of Algebraic Functions of One Variable. VI, 151 pages. 1973. DM 16,–

Vol. 315: K. Bichteler, Integration Theory (with Special Attention to Vector Measures). VI, 357 pages. 1973. DM 26,–

Vol. 316: Symposium on Non-Well-Posed Problems and Logarithmic Convexity. Edited by R. J. Knops. V, 176 pages. 1973. DM 18,–

Vol. 317: Séminaire Bourbaki – vol. 1971/72. Exposés 400–417. IV, 361 pages. 1973. DM 26,–

Vol. 318: Recent Advances in Topological Dynamics. Edited by A. Beck, VIII, 285 pages. 1973. DM 24,–

Vol. 319: Conference on Group Theory. Edited by R. W. Gatterdam and K. W. Weston. V, 188 pages. 1973. DM 18,–

Vol. 320: Modular Functions of One Variable I. Edited by W. Kuyk. V, 195 pages. 1973. DM 18,–

Vol. 321: Séminaire de Probabilités VII. Edité par P. A. Meyer. VI, 322 pages. 1973. DM 26,–

Vol. 322: Nonlinear Problems in the Physical Sciences and Biology. Edited by I. Stakgold, D. D. Joseph and D. H. Sattinger. VIII, 357 pages. 1973. DM 26,–

Vol. 323: J. L. Lions, Perturbations Singulières dans les Problèmes aux Limites et en Contrôle Optimal. XII, 645 pages. 1973. DM 42,–

Vol. 324: K. Kreith, Oscillation Theory. VI, 109 pages. 1973. DM 16,–

Vol. 325: Ch.-Ch. Chou, La Transformation de Fourier Complexe et L'Equation de Convolution. IX, 137 pages. 1973. DM 16,–

Vol. 326: A. Robert, Elliptic Curves. VIII, 264 pages. 1973. DM 22,–

Vol. 327: E. Matlis, 1-Dimensional Cohen-Macaulay Rings. XII, 157 pages. 1973. DM 18,–

Vol. 328: J. R. Büchi and D. Siefkes, The Monadic Second Order Theory of All Countable Ordinals. VI, 217 pages. 1973. DM 20,–

Vol. 329: W. Trebels, Multipliers for (C, α)-Bounded Fourier Expansions in Banach Spaces and Approximation Theory. VII, 103 pages. 1973. DM 16,–

Vol. 330: Proceedings of the Second Japan-USSR Symposium on Probability Theory. Edited by G. Maruyama and Yu. V. Prokhorov. VI, 550 pages. 1973. DM 36,–

Vol. 331: Summer School on Topological Vector Spaces. Edited by L. Waelbroeck. VI, 226 pages. 1973. DM 20,–

Vol. 332: Séminaire Pierre Lelong (Analyse) Année 1971-1972. V, 131 pages. 1973. DM 16,–

Vol. 333: Numerische, insbesondere approximationstheoretische Behandlung von Funktionalgleichungen. Herausgegeben von R. Ansorge und W. Törnig. VI, 296 Seiten. 1973. DM 24,–

Vol. 334: F. Schweiger, The Metrical Theory of Jacobi-Perron Algorithm. V, 111 pages. 1973. DM 16,–

Vol. 335: H. Huck, R. Roitzsch, U. Simon, W. Vortisch, R. Walden, B. Wegner und W. Wendland, Beweismethoden der Differentialgeometrie im Großen. IX, 159 Seiten. 1973. DM 18,–

Vol. 336: L'Analyse Harmonique dans le Domaine Complexe. Edité par E. J. Akutowicz. VIII, 169 pages. 1973. DM 18,–

Vol. 337: Cambridge Summer School in Mathematical Logic. Edited by A. R. D. Mathias and H. Rogers. IX, 660 pages. 1973. DM 42,–

Vol. 338: J. Lindenstrauss and L. Tzafriri, Classical Banach Spaces. IX, 243 pages. 1973. DM 22,–

Vol. 339: G. Kempf, F. Knudsen, D. Mumford and B. Saint-Donat, Toroidal Embeddings I. VIII, 209 pages. 1973. DM 20,–

Vol. 340: Groupes de Monodromie en Géométrie Algébrique. (SGA 7 II). Par P. Deligne et N. Katz. X, 438 pages. 1973. DM 40,–

Vol. 341: Algebraic K-Theory I, Higher K-Theories. Edited by H. Bass. XV, 335 pages. 1973. DM 26,–

Vol. 342: Algebraic K-Theory II, "Classical" Algebraic K-Theory, and Connections with Arithmetic. Edited by H. Bass. XV, 527 pages. 1973. DM 36,–